American Hegemony and World Oil

To my mother, Sid and Nan

American Hegemony and World Oil:

The Industry, the State System and the World Economy

Simon Bromley

The Pennsylvania State
University Press

Copyright © Simon Bromley 1991

First published 1991 in the United States by
The Pennsylvania State University Press, Suite C,
820 North University Drive, University Park, PA 16802

ISBN 0-271-00746-X

Library of Congress Cataloging-in-Publication Data

Bromley, Simon
  American hegemony and world oil / Simon Bromley.
    p.   cm.
  Includes bibliographical references and index.
  ISBN 0–271–00746–X
  1. Petroleum, industry and trade—Political aspects. 2. World
politics—1945–   3. United States—Foreign relations—1945–
I. Title.
HD9560.6.B68   1991
338.2'7282—dc20   90–47261
                                                      CIP

Typeset in 10 on 12 pt Times
by Photo·graphics, Honiton, Devon
Printed in Great Britain by
Billings & Sons Ltd, Worcester

11-23-93

Contents

Preface

***

> Security and economic considerations are inevitably linked and energy
> cannot be separated from either.
>   Richard Nixon at the Washington Energy Conference of 1974

The above notion serves as a motif which runs throughout this study.
Indeed, in working on questions of international politics, and specifically on the role of world oil, I too became increasingly disenchanted
with the separation of questions of political economy from those of
geopolitics – if for rather different reasons than President Nixon. Turning from political economy to the new historical, political sociology,
and thence to international relations, it was surprising to find that while
there existed a number of detailed studies of considerable insight, there
were precious few treatments which sought to focus simultaneously the
clearly related issues of geopolitics and economics. Through a case
study relating to world oil and the postwar hegemony of the United
States, this book is an attempt to indicate what might be involved in
such a project. The result differs in both its theoretical recommendations and its substantive conclusions from the mainstream assessments
currently uppermost on both sides of the Atlantic. Still, Richard Nixon
never let contemporary fashion stand in his way.

At this point, however, the disgraced president and I must part
company. Save for those able to see progress in a litany of defeats, the
period since the late 1970s has been a dark time for socialists. The
spread of inequality throughout the world, compounded by a wholesale
attack on the principles of planned social development, has brought a
recrudescence of right wing liberalism in the West and massive

deprivation in the Third World. The fillip given to this by events in the Eastern bloc is as yet far from clear. And while movements of workers and other popular forces have achieved some notable advances during the 1980s, the retreat of socialist politics has been accompanied by and reflected in a lamentable dispersal of radical intellectual work. In its place has arisen a timed conformism which basks in its self-professed, hard-headed realism.

The politics of this stance, together with the direct policy guidance offered – for example, that the advanced states should reduce their involvement with the underdeveloped world, or that they need to create a series of regimes in order to manage the common affairs of the West, or even that the political leadership in the United States should surmount the 'Vietnam syndrome' and once again exercise world leadership – betray an overriding managerial concern with order whose filiations to the new, conservative politics of inequality are all too obvious. But as in social theory, so in politics, the problem of order as formulated in broadly Hobbesian terms is neither a coherent nor a morally generous stance from which to address the problems of public life. In what follows I hope to contribute to the reshaping of a coherent radical challenge to the current academic orthodoxies. For, talk of a new epoch of capitalist growth and stability notwithstanding, the international system continues to face deep social crises, and radical analyses of these remain as urgent as ever.

Throughout the work for this book I have received a helpful admixture of advice, criticism, encouragement and libations from a number of people. Bob Jessop initially suggested the research topic out of which this came and has since shaped my thinking in many ways, and Geoffrey Ingham acted as a model supervisor for the original project. Kevin Bonnett and Tom Ling have also provided much needed encouragement during the dark night of Thatcherism. Justin Rosenberg not only served as a tireless sounding-board but also as a comradely exemplar of critical assistance, suggesting many detailed points for clarification and elaboration. David Held and Debbie Seymour of Polity Press have been patient and helpful – to them many thanks. Diyan Leake did an excellent job of copy-editing. My brothers, Paul and Philip, and sister, Sarah, have assisted my imbibitions on occasions too numerous to recall. Last but by no means least, Vanessa Fox has been a constant source of love and support throughout.

Simon Bromley, Sheffield

Introduction:
The Basic Argument

Following the dramatic events of the 1970s – including the momentous changes that transformed the world oil industry, the subsequent and apparently related economic crises, and the escalation of US military expenditure during the renewal of superpower confrontation in the Second or New Cold War – the 1980s witnessed a growing concern as to the longevity and durability of the global leadership of the United States. Was the era of US hegemony over, fallen victim of an inevitable imperial overstretch? By the end of the decade, the debate seemed to be closed: protagonists on all sides agreed that the real or imagined Soviet challenge of the 1970s had (for the foreseeable future at any rate) dwindled, yet the legacy bequeathed by the Reagan administrations to the US alliances with its North Atlantic Treaty Organization (NATO) allies in Western Europe, on the one hand, and Japan and the other Pacific powers, on the other, seemed to be altogether less profitable. In general, most commentators accepted that the US postwar hegemony was in more or less rapid decline, at least in the economic domain. And as for its geopolitical reach, since military power ultimately rested on domestic economic vitality, so the world influence of the United States would in time be similarly eroded.

Analysts differed primarily over the likely consequences of this loss of power by the United States in relation to the European Community and to the Pacific, thus opening up an extended policy debate around the most suitable remedial action. Realist theorists argue that an increasingly regionalized world economy, centred on the trading blocs of North America, Western Europe and the Pacific, will displace the open, liberal order that once secured postwar growth as well as Western unity in the containment of communism.[1] Further, now that communism no longer requires containing but is simply collapsing, the limits to US power will be starkly circumscribed by the emergence of dangerous currents over which it can have little control – a prospect only in part mitigated by the increased scope for intervention in the Third World. A more optimistic assessment for the post-hegemonic, post-Cold War world is advanced by writers of a liberal, transnationalist persuasion.[2] In this case, it is suggested that the Western powers will find institutionalized forms of co-operation through imaginative political leadership, constructing a range of international regimes to deal with common problems, while the collapse of communism signals not so much uncertainty as the world-wide triumph of liberal capitalist ideas.

Whatever the nature of such policy-oriented differences, these debates are structured and limited by certain shared premises: namely, a combination of realist protocols for the interpretation of world politics and of liberal or mercantilist versions of economic theory. Read through these optics, the relative material power of the United States has undoubtedly declined since the end of the Second World War – if not as rapidly as is conventionally alleged. For realists, US economic power is catalogued in such terms as the economy's share of world output, proportion of international trade, and productivity levels – similar measures are adopted for relative military power. Transnationalists consider, in addition, the role of the dollar in the world trade, as an instrument of finance and as a source of reserves, as well as the weight of US oil production in the global total. And though the dollar's role remains dominant, if currently underwritten by Japanese surpluses, the North American (US *and* Canadian) share of the world's oil production fell dramatically in the postwar decades, from 65 per cent in 1945 to 18 per cent in 1988.

By challenging the realist and neoclassical theories on which these accounts of world politics are based, the argument here contests both the theoretical basis of this consensus and the substantive judgements thereby licensed. Much has changed, to be sure, but neither the reach of US hegemony nor the suffusion of social conflicts across the skein of international politics is by any means over. The United States has lost some but by no measure all of the material preponderance it enjoyed at the end of the Second World War, while the structural reach of its power has increased and remains qualitatively and quantitatively more extensive than that of its competitors. And although the Cold War, in the form taken since 1945, may be coming to an end, the scope for revolutionary social change or violent ethnic conflict in large swathes of the world remains considerable. That these wider or structural features of the global system are rarely discussed is a testament to the hold of the realist and liberal orthodoxy over the study of international relations.

Indeed, it is no exaggeration to say that mainstream accounts of international politics have virtually ignored the dimensions of structural power in the global system, and that political sociology and economics remain trapped in a problematic of bounded societies, even when exogenous sources of change are considered. Specifically, the primary attention accorded to bounded nation-states and national economies pre-empts any serious study of the structures of the modern international system, thereby precluding an analysis of the structural power which shapes the behaviour of states and economic actors. In other words, the self-imposed limitation of international analysis to questions of what Susan Strange has termed the relational power between discrete agencies or institutions has vitiated the development of an adequate theory of world politics, to the particular detriment of a satisfactory account of US hegemony.[3] Now, as I argue in detail in chapters 1 and 2, the core structures of the international system comprise the capitalist world economy and the global community of nation-states, together with the associated world orders of military power and informational flows. In order to analyse the behaviour and power of a given institution or set of agencies – say, the US state or the world oil industry – it is necessary to locate its position and role within these structures, thus clarifying the historically changing forms of global power.

Paradoxically, the difficulties arising from the absence of such structural considerations in orthodox accounts can be seen most clearly in the area where it is generally claimed that the erosion of hegemony is most advanced, control over world oil. According to a leading proponent of hegemonic theory, Robert Keohane, it is in the degree of its control over global oil that the loss of US hegemony can be most visibly charted. Along with Stephen Krasner,[4] Keohane claims that US hegemony over world oil has been undermined by the actions of the producer-states in the guise of the Organization of Petroleum Exporting Countries (OPEC). However, in parallel with the divergent assessments of the scope for post-hegemonic co-operation sketched above, the agreement ends there: from the realist standpoint advocated by Krasner, the United States and with it the West is urged to reduce its dependence on a resource it no longer controls, whereas Keohane argues for the elaboration of an international regime – at least among the consumer nations – to replace the stability once provided by the hegemon. In reality, of course, public policy included a mixture of each strategy: by a variety of means attempts were made to reduce the West's dependence on Middle East oil and the International Energy

4   INTRODUCTION: THE BASIC ARGUMENT

Agency tried to serve as a regime for the consuming nations.

But neither of these proved to be the most durable or the most significant of policy shifts. Rather, the central response of the United States lay in a composite strategy embracing (1) the leverage deriving from its strategic and economic role in the Middle East, (2a) the manipulation of the privileged, structural position of the dollar in the world market, and (2b) the relative freedom of manoeuvre given by the asymmetric energy-independence of the North American bloc as compared with Western Europe and the Pacific. Also of great importance, though less obvious, has been (3) the permissive advantage resulting from the failure of the Soviet Union to use its huge petroleum resources to global political advantage. Indeed, what the Soviet experience suggests, in fact, is that the directive role of the United States in the world oil industry comes not only from its reserves but also from its structurally privileged position in the postwar world economy. Put differently, the *differentia specifica* of US hegemony lies in the capitalist structure of the world system as much as in the institutional preponderance of the United States.

In order to address these issues I develop a *conjunctural* model of world politics – that is, one in which the structural features of the world system are explicitly theorized. The vexed question of US hegemony in the domain of oil provides the case study around which the discussion is focused. The aim of this book, then, is to provide an alternative framework for thinking about international politics by way of explicating the connections between world oil and US hegemony. Accordingly, reviewing the standard international relations literature and relevant developments elsewhere in the social sciences, chapter 1 both elaborates a conjunctural model for international relations theory and offers a cognate specification of global hegemony. Breaking with the reifications of realism, in which a substantive theory of the international (the balance of power) is mistaken for a general conceptual framework, it is suggested that all social analysis has to operate at three levels of abstraction in order to capture the nature of agency and power in, and hence the causal dynamics of, social and historical development. Substantive organizations and processes can be studied at the level of structures, institutions and actions. In particular, it is necessary to theorize the behaviour of the dominant institutions of world politics – the oil industry, for example – in terms of their shaping by the analytically distinct yet substantively interwoven structural parameters of the modern world.

Moreover, these parameters – such as the structure of sovereignty which delimits the nature of political power in the nation state system, or the structure of capital which shapes the character of economic power on a world scale – have their own temporalities, developing in complex relation to one another. In chapter 2, therefore, I develop the necessary historical dimensions of the conjunctural account by applying it both to the central problems in the geopolitical economy of the world oil industry and to the rise of US hegemony, thereby elucidating some of the connections between these two phenomena. This involves situating the emergence of US world leadership in the context of an expanding and consolidating nation-state system as well as in relation to the developing capitalist world economy. To highlight the importance of oil and its structured role in the organization of US hegemony, I characterize it as a *strategic* commodity – significantly something it never became for the Soviet bloc. Chapter 3 extends this analysis, placing the rise of international oil in the context of the consolidation of the nation-state system consequent on the collapse of European empires, on the one hand, and the imperialist economic penetration of the periphery, on the other. A complementary narrative relates this process to the coincident and related shift from a European 'balance of power', marked by British global dominance, to an era of 'world politics', fractured by Cold War and managed under US hegemony.

On the basis of this analysis, chapter 4 directly challenges the conventional wisdom noted above on one of its key claims: namely, that the rise of OPEC undermined the company-centred, US-dominated world oil order and with it a vital prop of US hegemony. By focusing on the integration of the main Gulf states into the world market and their dependence on US military power, I show how a focus on OPEC *qua* organization, rather than on the focal positions of Saudi Arabia and Iran, can be seriously misleading. Further, a broader focus also suggests that the economic crises of the 1970s and early eighties were not the results of exogenous shocks delivered to a stable economic system but were the direct consequences of the necessarily uneven development of the world economy and the nation-state system. Without seeking to minimize the dramatic quality of the institutional changes in the organization of world oil which so dominated the early 1970s, together with the altered relations of power among its constituent actors, it is important none the less to recognize that the global power of the United States remained substantially intact. In turn, this offered significant levers of influence in the domain of oil. Thus, notwithstanding a diminution of US material preponderance, profound institutional change has not been inconsistent with the structural maintenance of US hegemony, albeit taking a new form in the altered conjuncture of the late 1980s.

6   INTRODUCTION: THE BASIC ARGUMENT

Chapter 5 extends this argument, considering the positions of Western Europe and Japan in the world oil order as well as the failure of the Soviet Union to provide an international alternative to Western dominance over Middle East oil. On the basis of a reunified world market, the postwar recovery and growth of Western Europe and Japan came to be heavily reliant on cheap oil, sourced by the (mainly US) majors from the Middle East. The geopolitical ramifications of this are then explored, demonstrating that the events of 1973–4 did signal a partial loss of US control over the energy policies of its allies, thereby compromising its leadership role. But as the events of 1987 in the Gulf indicated, the NATO allies eventually followed the US lead. On the other hand, the Soviet Union, with its steady expansion of petroleum exports to the West and its inability to construct a radical Arab coalition in the Middle East, proved no match for the United States. In fact, while the Soviet bloc achieved significant planned, extensive economic growth, the qualitative advantage of the world market, based upon its high degree of alienability of resources, proved a greater attraction for developing regimes, especially for those with secure capitalist classes. For the Soviet Union this created a relation of dependency wherein Soviet reform and food imports were in part tied to finance generated by hard-currency oil and gas exports. In sum, while the dramatic shift in the balance of advantage between East and West during the 1980s significantly altered the US relations with Western Europe and Japan, throughout the post-OPEC developments the superpower relationship remained the central point of reference for European and Japanese overtures either to Soviet energy supplies or to the Middle East.

Finally, in chapter 6 I outline the changes behind the geopolitical economy of Reaganism, the transformations in the oil industry, and the reshaped position of the United States in the Middle East, showing how its directive role in world oil was reconsolidated in the 1980s, if in an increasingly unilateral and predatory form. By the late 1980s it was apparent that the power of the oil producers was limited when considered against the backdrop of their economic and military ties to the West and the United States in particular. For, despite a range of untoward developments in the Middle East, the United States was able to rebuild a regional alignment that has served as the basis for a renewed integration of the Gulf producers and Western transnational corporations, thereby supporting long-range price stability and an unambiguous pro-Western orientation. Moreover, the structural power of the United States – above all, the continued dominance of US corporations in world energy markets, the relative energy self-sufficiency of North America, and the dollar denomination of most world oil trade – in the sphere of energy, to say nothing of its broader economic and military power, starkly underscores its continuing position of dominance in this key domain. Of course, nothing can guarantee against a regional upheaval which would compromise this new stability – but then no hegemonic power has yet found a way of forestalling revolutionary social transformations.

1
Rethinking International Relations Theory: A Conjunctural Approach

Introduction

A prominent feature of international relations as a discipline, and one which continues to be reproduced in its search for general theories of international politics, has been the effective division of the subject into strategic studies and the politics of international economic relations.[1] This is reflected at a theoretical level in the separation between the realist discourse of national security and the study of international political economy. As a result, within what is commonly taken to be a potentially unified theoretical field, this disciplinary and theoretical division of labour has considerably complicated the task of basic conceptual clarification. Furthermore, the relative intellectual distance between international relations theory in general and developments elsewhere in the social sciences, most notably in political sociology and social theory, has reinforced the marked lack of progress within the discipline.[2] Within the discipline itself the protracted debate between the (neo-)realists and the transnationalists reached no firm conclusions and is now at something of an impasse.[3] At the same time, a number of central debates in the social sciences are of direct relevance to the concerns of international relations theory. In economics, sociology and geography a renewed interest in questions of international political economy has prompted a number of vigorous debates centred around a series of mercantilist, Marxist and ecological challenges to the dominant liberal paradigm.[4] This in turn has built upon the growing recognition of and desire to transcend the polarized and sterile debates between traditional modernization theory and neo-Marxist dependency theory.[5]

Meanwhile, in political sociology there has recently been something of a rediscovery of the state. This attempt to 'bring the state back in' focuses on the theme state autonomy and the related question of the degree to which the behaviour of nation-states, whether in the domestic or international arena, can be said to be independent from the economies and societies over which they preside.[6] A closely connected development in sociology and social theory is the current attempt to theorize the distinctive social role of militarism,[7] together with the related if distinct project of understanding the administrative power of the modern nation-state.[8] Finally, at a more general level of analysis, the 1970s and early eighties were preoccupied with a range of conceptual and methodological debates in which the basic questions of the relations between structure, agency and power in social theory were addressed.[9] Of course, the evolution of these debates has been markedly uneven and the potential connections between each of them have rarely been drawn. But it is increasingly clear that theoretical progress in the field of international politics and economics requires a high degree of cross-fertilization among the social sciences. And nowhere is this more true than in the field of mainstream international relations theory.

Accordingly, in this chapter I propose to consider (1) the current state of international relations theory, paying particular attention to the theoretical underpinnings of the debate between realism and transnationalism as well as the main trends in international political economy, (2) the nature of the disciplinary impasse reached, especially as demonstrated in the debate around Waltz's attempt to provide a systematic formulation of realism, and (3) the manner in which some of the controversies in the other social sciences noted above might begin to provide a more adequate theoretical basis for the study of global politics and economics.

International Relations Theory: Fracture and Convergence

The Development of the Discipline

The standard accounts of the development of the discipline of international relations, which in turn accurately mirror its self-image, tend to portray it in terms of the supersession of idealism by realism – a process that is itself viewed as one of decisive cognitive advance.[10] Thus, it is suggested that before the First World War the study of relations between states was parcelled out among law, history and philosophy, and that it was only with the interwar era that a self-sufficient discipline of international relations emerged from its phase of youthful exuberance. In this account, idealism is regarded as having dominated the subject between the wars, and is characterized by its attempt to develop methods of peaceful crisis resolution through the codification of international law and the spread of international organizations. In the light of the interwar slump, the drift to global war and the multiplying national hatreds of those years it is then all too easy to argue that idealism was simply a reflex of both the antipathy to war generated by the experience of 1914–18 and the ideology of the status quo powers. It was, in other words, of a piece with the optimistic liberal internationalism that lay behind (at least the rhetoric of) the League of Nations and Wilson's Fourteen Points.

From this vantage point, Carr's seminal work *The Twenty Years' Crisis*, and, after the war, Morgenthau's *Politics among Nations* are taken together to have placed the study of international relations on a new, scientific footing.[11] Realism was hereafter to be concerned with the basic questions of national 'interest', 'power' and 'security'. Although Carr pronounced 'the realist critique of utopianism', these terms denoted for him not successive formal schools of theory but antithetical normative and critical modes of human thought, mutually necessary in the formulation of 'sound political thought and sound political life'.[12] It was Morgenthau who, in a markedly cruder manner, asserted that 'the history of modern political thought is the story of a contest between two schools that differ fundamentally in their conceptions of the nature of man, society, and politics'[13] – and who then deployed this contrast to distinguish a self-sufficient realist theory of international politics. The six principles of political realism, which Morgenthau adduces, are meant to provide the basis for a scientific study of international politics. In essence, Morgenthau's argument reduces to two related claims: that 'politics, like society in general, is governed by objective laws' which are rooted in an unchanging 'human nature';[14] and that international politics can be understood through 'the concept of interest defined in terms of power' – a conception which he claims delimits 'politics as an autonomous sphere of action and understanding'.[15]

Now while these texts are conventionally taken to constitute a realist break with the idealist or utopian past and thus to found the basis of a science of international politics, their true significance is that they license a basic framework which has since set the boundaries of realism. These boundaries are delimited by, on the one hand, the practical discourse of *realpolitik* and its associated concepts of sovereignty, anarchy, power and interest, and, on the other, the positivist self-understanding of the discipline. Indeed, because realism is defined less by its possession of a common body of theory than by such assumptions, some have questioned whether there is in fact a single and coherent realist tradition. For example, a distinction is often drawn between the clear differences of emphasis among the European (or, more accurately, British) and the US traditions of realist thought.[16] Each of these shares the supposition that sovereign states endure in a condition of international anarchy, yet the former stresses that a degree of international *society* might emerge on the basis of a stable balance of power.

Perhaps in part because of this lack of internal coherence but also as a consequence of the obvious failure of realism to develop theoretically, critiques of the discipline proliferated during the 1960s and seventies. Equally important, historically speaking, was the growing concern with questions of economic interdependence among the leading Western states, together with the apparent irrelevance of realist suppositions to many of the internal relations of the Atlantic alliance, which produced a growing challenge to the dominance of the realist discourse of national security analysis and prescription.[17] In the course of this challenge, the classical realist assumptions were progressively relaxed in a search for an ever more general theory of international relations. This critique of realism debouched into a transnationalist model of complex interdependence and an increasing focus on the analysis of international regimes.[18]

A parallel trajectory characterized the evolution of debates within the sub-discipline of international political economy.[19] In this case, the erstwhile dominance of liberal economics was increasingly challenged by the revival of mercantilist perspectives and the resurgence of Marxian ones. As the stable monetary and trading order of the long-boom appeared to break down, liberal political economy also came to focus on the presence of absence of international co-operation organized through regimes. Another signal of this partial convergence was the attempt by some to integrate a modern version of mercantilist economics into a basic political framework derived from realism.[20] This realist–mercantilist political economy of international relations still regarded the question of *state power* as the primary focus of attention, but it offered much broader notions of *power* and *security* than were commonly invoked by the politicomilitary focus on strategic advantage, military preparedness and the maintenance of territorial and political integrity.

In the mean time a realist counterattack was developing. Here it was, above all, Kenneth Waltz whose work explicitly sought to reformulate and *theorize* the boundary assumptions of classical realism, thereby developing a pure theory of *international* politics through a series of deductions from the principle of anarchy.[21] And it is a cardinal virtue of Waltz's work to have demonstrated what is involved in defending a strict version of realism. For if realism is broadened to include an admixture of mercantilist and liberal economic theory (what Waltz would term a study of unit-level processes), then the path towards some form of compromise with the claims of transnationalism becomes very slippery indeed. As a consequence, some recent critics have sought to excise Waltz from the realist tradition by distinguishing between an original, classical and critical realist tradition and a later, narrower, problem-solving science of neorealism.[22] At first, many saw the issues involved in these debates as essentially empirical. It soon became clear, however, that substantive theoretical disputes were involved. This can be seen most clearly in the form taken by the realist counterattack and the subsequent debate between neorealism and its critics. For here the original, empirically based transnationalist critique of realism was reproduced in the form of a number of theoretical objections to Waltz's systems theory model of realism.

Classical Realism and the Transnationalist Critique

The textbook summaries of realism typically comprise four basic propositions.[23] First, the international order comprises a system in which the pre-eminent actors are *states*. Second, because of the inherently anarchical nature of international society, the fundamental issue of concern to states and the dominant ordering principle of international politics is state *security*. Third, in order to guarantee their security states must seek a favourable *power* position, and this can be achieved either through national measures or through alliances. Fourth, it is maintained both that the above assumptions accurately characterize the perceptions and actions of foreign policy elites, and that these elites are able to formulate such policy independently of any significant pressures from their domestic political and economic systems.

The transnationalist perspective, by contrast, has stressed the complex interdependence of the relations between national economies and states. Theoretically, it is derived by relaxing the strict realist axioms noted above in order to account for the political processes of an increasingly interdependent world. Indeed, transnationalism began from the observation that idealism and realism shared certain important state-centric assumptions as to the nature of the object of study in international relations. Namely, each subscribed to the view that nation-states are the key actors in international relations, that the focus of study should be the issues of power and security, and that a clear separation exists between domestic and foreign policy. It was these state-centric axioms that came under attack from the transnationalist school. Transnationalists argued that the postwar world had witnessed historic transformations of such scope that the object of international relations theory had been fundamentally altered. In particular, these authors drew attention to changes in the *structure* of the global system, alterations in the *processes* of international relations, and the emergence of new *actors* in international society. In short, if states were no longer the dominant actors in certain (economic) issue areas, if power (specifically military power) was not the primary instrument of policy and focus of study, and if there were other significant units in international society apart from autonomous states, then the state-centric paradigm of old had to be replaced.

More concretely, the breakup of the bipolar superpower order had resulted in a new international structure: each block was subject to pressures of internal fragmentation; secondary powers increased their economic and military capabilities; and new actors entered international society. The processes of world politics were transformed by the displacement of security concerns by economic interests: in the seventies it seemed to many that the overriding East–West cleavage in world politics was giving way to a North–South divide. And finally, the primacy of the nation-state as the sole object of international relations was increasingly compromised by the interactions of two developments: first, with the advance of national economic, political and military interdependence, the distinction between domestic and foreign policy was eroding; and second, with the emergence of new non-state actors (either transnational or supranational as in OPEC or the European Community), the centrality previously accorded to states appeared to be misplaced – especially if economic issues were displacing those of security.

The clearest statement of the new paradigm of complex interdependence came in the late seventies from Keohane and Nye's *Power and Interdependence*.[24] In contrast to the realist assumptions detailed above, these authors argued that a state of affairs characterized by complex interdependence had three main features.[25] First, because of the intrusion of economic issues into international politics there was also no clear separation to be drawn between domestic and international politics; and one had, therefore, to study – in addition to the state-to-state linkages studied by realism – the role of transgovernmental relations and organizations and the transnational linkages formed by non-state actors pursuing their own private interests. Second, within certain regions and in particular 'issue-areas' the deployment of military power was no longer a realistic option for the state. Third, this increased salience of the economic and reduced efficacy of the military dimensions in world politics meant that there was no longer any clearly defined hierarchy of issues or uniform institutional arena (the state) in which questions of security could be said to predominate. On this account, the conditions of complex interdependence and so the applicability of transnationalism are likely to increase in the modern world because they reflect long-term though not necessarily irreversible developments: the absence of a clear hierarchy among issues reflects the intrusion of welfare concerns into international politics; the complexity of interstate linkages results from the expansion of global transport and communications; and the reduced salience of military power derives from the combination of the increased destructiveness of modern weaponry and the increased potential for social and national mobilization in Third World states.

Because the object of international relations theory was characterized by an incipient shift from international anarchy toward complex interdependence, the paradigms of realism and transnationalism are best conceived of as ideal types. But Keohane and Nye nevertheless maintained that 'our three conditions [for complex interdependence] are fairly well approximated on some global issues of economic and ecological interdependence and that they come close to characterizing the entire relationship between some countries'.[26] One important consequence of this powerful challenge to the realist premises was that these features identified by transnationalism gave 'rise to distinctive political processes, which translate power resources into power as control over outcomes. [And so] . . . under conditions of complex interdependence the translation will be different than under realist conditions'.[27] Thus, each of the following is viewed differently under realist and transnationalist assumptions: the goals of foreign policy, the instruments of international politics, the process of agenda formation, the degree and character of linkages between issues, and the role of international organizations.

Complex interdependence was also associated with the growing importance of international organizations and regimes: regimes and their associated organizations constituted the emergent institutional form of complex interdependence, which had come to overlay the (realist) balance of power. In this context Keohane and Nye pose the question, 'Why do international regimes change?' and suggest that the answer be found in 'four models based respectively on changes in (1) economic processes, (2) the overall power structure in the world, (3) the power structure within issue areas, and (4) power capabilities as affected by international organization'.[28] In general, they predict and find that the closer reality approximates to a condition of complex interdependence, so the better a power structure within issue areas and/ or an international organization model will explain outcomes. Similar considerations apply in the case of bilateral relations between states. These can be characterized by the extent to which complex interdependence pertains between the states concerned together with the degree of asymmetry involved in the relationship. And again, to the extent that complex interdependence and symmetry obtain, so transnationalism results in a better explanation of outcomes than realism.

In this way international relations theory reoriented itself away from the narrow focus of realism towards a consideration of economic processes and a relaxation of the unitary model of the state. In a similar manner, the study of international economic relations moved away from the pure theory of trade and development towards a focus on the political regulation of the global economy. Thus, not only was the study of international relations in general considerably enriched but also there developed something of a double convergence: one wing comprised the advocates or regimes analysis (the transnationalists and the liberal economists); and another grouping was constituted by the attempt to develop a synthesis of mercantilism and realism. In economics, this shift towards a concern with regimes and politics was registered in the revitalization of the tradition of political economy.

International Political Economy

Liberal accounts of the international economic order have largely been an extension of the basic theoretical assumptions of neoclassical economics to the relations of trade, money and capital flows between national economies. Formed in the era when the social sciences sought scientific (and indeed political) respectability through the importation of mathematical formalism together with a rejection of historical generalization in favour of empiricism, orthodox economics – pioneered in the work of Jevons, Walras and Menger – constituted a sharp break with the perspectives of the classicals, Smith, Ricardo and their greatest critic, Marx. Broadly speaking, orthodox economics has since been characterized by the prominence of a theory of resource allocation based on marginalist forms of analysis, in which maximizing behaviour

is brought into equilibrium through the market, and in which each
factor of production has 'theoretical parity'.[29] By their rational pursuit
of utility maximization agents are effectively constrained within this
framework, and so outcomes are strongly determined. The methodology of economics is construed in logical empiricist terms: economics
is a 'positive science'. Now, while this marginal revolution in economic
thought considerably expanded the scope of formal analysis, and incidentally was coincident with the professionalization of the subject, it
simultaneously narrowed the focus of study: 'This meant that questions
of historical dynamics, of economic development and indeed of economic fluctuations and crises were largely extruded from the fields of
the new academic orthodoxy'.[30]

A second turning point in the development of mainstream economics,
the emergence of macroeconomics as a distinct branch, came with the
appearance of Keynes's *General Theory* (1936), though the postwar
period also witnessed the formal statement of general competitive
analysis (Arrow–Debreu) along with the emergence of econometrics as
reliable statistical data accumulated in tandem with increased state
intervention. Arising from this Keynesian *démarche*, there have been
a number of macroeconomic analyses which focus on the interrelations
in the world economy: here each national economy is described in
terms of a set of aggregate indicators – investment, savings, government
spending, taxes, imports, exports and foreign borrowing – and then the
compatibilities and disjunctures between different national economic
configurations are examined. Along with the discussion surrounding
Leontief's apparent refutation of the Heckscher–Olin theory of trade,
the application of Keynesian models to the theory of the balance of
payments opened new questions as to the nature of international trade
and, in particular, its benefits for developing countries. Development
economics, itself unheard of before the 1940s, originated in this environment as a host of newly independent states in the Third World sought
to industrialize.

In both neoclassical microeconomics and Keynesian macroeconomics
the market-ordering of economic relations is held to increase economic
efficiency, maximize growth, promote consumer welfare and underpin
individual freedom. In this perspective, the forces of supply and demand
determine the relative prices of goods and services; and, in turn,
these prices move to allocate resources, organize production and clear
markets. With the absence of 'exogenous' obstacles to the functioning
of 'free markets', the effect is that the rational pursuit of utility under
conditions of scarcity will result in a socially optimal outcome of economic equilibrium – where both societies' resources are efficiently

employed and individual welfare is maximized. Of course, this picture is something of a caricature – but it is not seriously misleading. For although the application of liberal political economy to the international arena recognizes that these assumptions are highly abstract, it is none the less maintained that they do capture the basic animating forces of economic life.

More formally, the liberal perspective is theoretically grounded in the general equilibrium model of market economies.[31] The analytical procedures of this model are premissed on an individualistic methodological stance combined with the argument that the animating force of economic activity resides in the rational processes of utility maximization by discrete, egoistic and non-satiated agents. Economics is, then, 'the science which studies human behaviour as a relationship between ends and scarce means which have alternative uses'.[32] The analysis takes utility functions, technology, and initial endowments as given and constructs a series of supply and demand curves which when solved in mechanical (reversible) or logical (causal sequence) time describe the equilibrium state of the economy. The simultaneous solutions of these equations determine the prices and quantities compatible with market clearing behaviour. This mode of analysis is inclusive, the purpose is to treat as many variables as possible as endogenous (determined within the theoretical framework), and the locus of explanation resides in those variables which remain exogenous.[33] As to the precise relation between the pure micro-theory and the more widely practised macroeconomics, the current preoccupation of much mainstream economics is concerned with the attempt, by way of summing the decisions of discrete agents into additive aggregates, to ground explanations of macro-behaviour in micro-theoretical assumptions.[34]

A number of critical limitations beset this model. Two important formal requirements for the equations of general equilibrium theory to have a determinate solution are that each agent in the economy is a price-taker and that perfect foresight is attainable through the existence of a full set of futures and insurance markets. In turn, an important effect of these requirements is to exclude a theory of power: since all agents conform to market signals none can alter the parameters of the market. Equally, a conception of historical (or irreversible) time is difficult to incorporate into general equilibrium theory as its very presence is assumed away in the notion that futures markets provide insurance against uncertainty.[35] The micro–macro transition is also deeply problematic in so far as it provides no means of accounting for the structuring of the decisions of agents by economic wholes. For, as Hodgson argues, 'wholes cannot be compared to additive aggregates at all . . . In summations the parts function because of their inherent qualities. On the other hand, when a number of parts constitute a whole, the parts do not enter into such a connexion by means of their inherent qualities, but by means of their position and function in the system'.[36] Another closely related weakness deriving from the attachment to a strict methodological individualism is that 'the influence of social institutions on individual goals and choices is almost entirely excluded by mainstream economic theory'.[37] A final problem is of an epistemological kind: namely, the theory of general equilibrium defines its object (the market economy) in terms of its own assumptions and in consequence the theory reveals 'little other than its own [logical or axiomatic] structure'.[38]

In sharp contrast, Marxist perspectives in international political economy stress the role of power and the collective (class) pursuit of sectional interests. The collective agencies are taken to be social classes defined in terms of their relations to the means of production. Combined with its analysis of the class nature of the state, this focus has provided Marxism with a number of theories of international relations. Marx himself wrote much on international issues – especially in his journalism of the 1850s but also in *Capital* and elsewhere – but as is well know he left nothing of the theoretical power of *Capital*. None the less, the core of Marx's writings on the 'international' comprised a linked insistence on the social character of the state form, the transnational scope of the capitalist mode of production, the combined and uneven nature of all social development and the competitive struggle of states seen as embodying antagonistic modes of production. And, notwithstanding the widespread misapprehension that Marxism and international relations theory have little to say to one another, these themes can be found in the classical debate on imperialism associated with Hilferding, Bukharin and Lenin.[39] Equally, the related if differently focused arguments of Luxemburg and (perhaps above all) Trotsky were explicitly concerned with theorizing the transformations wrought on social development and international relations by the increasingly global consolidation of capitalism. Together this body of writing provided classical Marxism with a more developed theory of international politics than can be found in the writings of Marx and Engels.

Lenin's discussion of *Imperialism* (1917), the increased militarization of the leading capitalist powers, and the support of the European workers' movements for the First World War is the most celebrated if not original contribution to this debate. Lenin's pamphlet was but a 'popular outline' of the argument developed in Bukharin's *Imperialism and World Economy*: that inter-imperialist rivalry, which was the result of the conjunction of the internationalization of capital with the formation of finance capital (as analysed by Hilferding), constituted the dominant level of world politics.[40] According to Lenin's analysis, imperialist rivalry explained the massive expansion of formal colonialism between 1880 and 1914 as well as the origins of the Great War. The competitive, global spread of monopoly capital, which together with the rise of substantive state intervention defined imperialism, could take the form of international trade, the export of capital and the development of multinational corporations. Now, the details of this argument have been rightly criticized: on the empirical grounds that capital export was not as important for the key capitalist powers as Lenin suggested, and that which did take place was primarily intrametropolitan rather than oriented towards the scramble for colonies; and the stagnationist theory of monopoly capitalism, beset by worsening crises of underconsumption, failed to consider the full extent of its recuperative strengths.

What is perhaps more important, however, was that Lenin (following Hilferding and Bukharin) drew attention to the advent of a new phenomenon – namely, the development of a qualitatively distinct stage of capitalism and, associated with this, a transformation of the role of the state both domestically and internationally, which therefore changed the terms of international rivalry.[41] At a substantive level, then, Lenin's pamphlet effected a decisive reorientation of attention away from the domestic politics of the *classes* in the core capitalist economies, and towards the global impact of capitalism on its periphery and the militaristic and mercantilist policies of the core *states* in their relations both with each other and to the exploited periphery. This gave classical Marxism a theory of *international* politics: the primary actors in global politics became competing mercantilist and militarist nation-states driven by increasingly global economic forces.[42] Lenin's work is only the best known contribution to the classical Marxist debate over the character of imperialism, a debate that Halliday has characterized as 'an attempt to theorize the relationship between socio-political system, arms production and the expansion of the international economy and at the same time to theorize the politico-economic character of international hierarchy, beyond mere assertions about unequal power'.[43]

Of equal importance for a Marxist theory of international politics was Trotsky's generalization of his theory of the scope for permanent revolution in Russia, which he adumbrated after the Russo-Japanese War (1904–5) and the subsequent Revolution in *Results and Prospects* (1906), to the possibilities for accumulating peripheral revolutions. As

Mandel notes: 'Underpinning this notion of world revolution is . . . the concept of world economy and the class struggle as a totality subject to uneven and combined development'.[44] Now, it is not necessary to accept either Trotsky's (and Mandel's) ascription of a tendential unity to the process of 'world revolution', or to regard the First World War as a turning point marking the demise of capitalism as a global system, to see that his analyses of the constraints on socialist development in backward countries, together with the antagonistic relations which would persist between these states and the advanced capitalist states, provide a host of insights into both the West's relation to the Soviet Union (and other socialist states) and the Cold War.[45]

Turning now to the formal properties of Marxian economics, these also differ radically from those of liberal, general equilibrium theory. In the surplus tradition capitalist economies are understood as class-divided systems of generalized commodity production and exchange. The animating forces of the economy are the imperatives that drive capital to accumulate under the constraints of class struggle and inter-capitalist competition. This model regards economic growth, or capital accumulation, as a fundamentally *dis*equilibrating process. The imperative of competition forces capital to revolutionize and expand the forces of production without regard to the parameters of the market, and hence the proportionalities – within given branches, among the branches of production, and between production and consumption – necessary for stable accumulation are constantly eroded by the overaccumulation and uneven development of capital. Marx's own labour theory of value and his account of the genesis and character of capitalist crises is one attempt to theorize these dynamics. Unfortunately, however, what Clarke has noted in discussing Marxist theories of crisis applies more generally:

> In general Marxist crisis theories have been concerned to prove or disprove the inevitability of crisis within an equilibrium theory, based on Marx's reproduction schemes or his analysis of the law of the tendency for the rate of profit to fall, rather than exploring the historical dynamics of overaccumulation and crisis within the kind of disequilibrium theory that dominates Marx's own work.[46]

Indeed, most contemporary mathematical treatments, with their search for an equilibrium theory of price determination, often obscure this central theme. Thus Sraffa's attempt to dispense with the labour theory of value, and thereby to formalize this model, notoriously dodges the problem of effective demand and says nothing about long-term growth. Some post-Keynesians, such as Joan Robinson, have concluded that equilibrium theorizing must therefore give way to an historical approach, but this stance simply converts the complexity of all real historical change into an argument for dispensing with formal models of theoretical determination. For the formal models can still clarify certain important economic relationships which any historically oriented account must attend to.

The other less well known formal model in the surplus tradition is that due to von Neumann. Von Neumann's model of the economy takes the real wage to be exogenous and accumulates all the surplus such that output grows at a constant rate without limit. Sraffa, on the other hand, derived an inverse relation between the real wage and the rate of profit, but considered output as given and said nothing determinate about the actual distribution of the surplus. Goodwin and Walker, among others, have shown how in principle these respective defects can be remedied in a more general dynamic model.[47] What is involved is the construction of a dynamic adjustment model in which investment varies (in quantity and composition) to accommodate, in part at least, capacity to demand. (It can only do so 'in part' because, in contrast to the assumptions of general equilibrium theory, the rate of technological change is largely endogenous in modern economies.) This model uses the Keynesian notion of the multiplier-accelerator in order to illuminate the character of cyclical dynamics, but differs from Keynesianism in both the degree of autonomy accorded to investment targeting and the plasticity accorded to production.[48] For Keynesians, as for those who adhere to the Kaleckian theory of monopoly capitalism, prices are set to generate necessary profits required to finance the desired level of investment, itself a function of demand. But, as Walker has pointed out, the theory of 'investment targeting needs to be anchored to the long-run growth trajectories of different sectors . . . and not left free-floating according to the spirits of various capitalists'.[49] The reason for this is that capitalists invest and compete in search of profits, rather than seeking to meet demand.

Overall, because of its focus on collectivities of economic agents, their competitive struggle to produce and distribute the surplus, and the specific determinants of investment decisions, the surplus tradition is better able to accommodate notions of power and real time into its analysis. Equally, because it is less concerned with pure equilibrium models than with the disequilibrating consequences of uncoordinated and concrete investment decisions, which are themselves specific to particular sectors, the surplus approach can in principle study the institutional and structural shaping of economic behaviour. Finally, the surplus approach is compatible with the realist epistemological stance of retroduction: this involves the identification of an *explananda*, then by a posteriori reasoning an *explanans* explains the *explananda* provided it exhaustively accounts for it, is itself internally consistent, and is logically independent of the terms of the *explananda*.

Economic nationalism or mercantilism can be dealt with more briefly. It has a long history in economic thought, taking several different forms from mercantilism through to the New Protectionism. Theoretically it stresses the role of power, conflict, and the collective (national) pursuit of sectional interests in the sphere of international economic relations. It is also attentive to the particular institutional and structural contexts of major investment decisions. In policy terms this tradition subordinates economic questions to the requirements of the state, and the primary interests of the state are identified as the joint pursuit of national security and military power. Viner has summarized economic nationalism in terms of

> the following propositions: (1) wealth is an absolutely essential means to power, whether for security or for aggression; (2) power is essential or valuable as a means to the acquisition or retention of wealth; (3) wealth and power are each proper ultimate ends of national policy; (4) there is a long-run harmony between these ends, although in particular circumstances it may be necessary for a time to make economic sacrifices in the interest of military security and therefore also of long-run prosperity.[50]

More recently, Sen's study, *The Military Origins of Industrialization and International Trade Rivalry*, has argued that state-led industrialization is the primary goal of national development.[51] The growth of industry introduces economies of scale that have an energizing effect throughout the economy, industrialization can lead to a degree of economic security for the nation-state, and finally the possession of a group of 'strategic industries' is essential to the military security of the state. In turn, military security is an essential precondition for political participation in an inherently anarchical inter-state system. This argument suggests that economic nationalism is not merely a response to nationalist ideologies of development in conditions of uneven economic growth, but is also an ineradicable feature of any development in the context of the modern nation-state system. Drawing on Keynesian theory, mercantilism also provides a theory of imperialism in so far as states seek to secure control of foreign markets to guarantee demand as well as to gain access to raw materials to lower costs.

Mercantilism does not, however, possess a theoretical underpinning of its own, in the way liberal or Marxian approaches do. And neither does its theoretical form adequately reflect the changing articulation of national and international economies throughout the history of capitalist development, to say nothing of the discrete dynamics and logics of state socialist economies. In so far as economic nationalism seeks an analytical framework it has drawn on the neo-Ricardian appropriation of Sraffa's work. In this respect it is, therefore, closer to Marxian than liberal perspectives in its concentration on questions of power and the distribution of the economic surplus between social groups. In its focus on nationally unified economic units, however, it is closer to Keynesian versions of the liberal tradition than the transnational, class-based perspectives of Marxism.

Before moving on it is instructive to note some of the overlaps between traditional international relations and international political economy. First, both realism and economic nationalism stress the primacy of the interests of the state, and to some extent they both reflect upon different aspects of the predicament of the state in what is taken to be a basically anarchic order.[52] There are also similarities between a Leninist theory of international politics and realism:[53] both see states struggling for an economic surplus and military power, and each regards the uneven development of power on a global scale as the key dynamic of international politics. However, in this case there are also important differences: for Lenin it was ultimately (transnational) economic forces which provided the motor of international relations, whereas for the realists an analytic primacy is accorded to the international imperatives faced by states in the fields of security and prestige; of equal importance, Lenin's notion of the state as a set of class-based, coercive institutions differs radically from that implicit in the realist conflation of government, state and nation.[54]

For these reasons it is an error to assimilate the classical Marxist theory of imperialism to realist–mercantilist accounts. As Barratt Brown explains:

> Colonialism for the Keynesians [and this applies equally to realism and mercantilism] has to do with nation-state measures to expand exports of goods, capital and labour and to increase the cost of unwanted imports. Imperialism for the Marxists has to do with capitalist firms seeking for surpluses and seeking to use their surpluses wherever they can by incorporating new areas of the world economy into their system of accumulation.[55]

Another significant connection might be suggested between transnationalism and liberal theories of international economic interdependence. Both of these perspectives stress the need for interstate co-operation and compromise in a world of complex interdependence, and each regards interdependence as leading to the displacement of antagonistic security issues by the fundamentally peaceful questions of economic management. Finally, as noted above, the transnationalist critique of realism and the revival of political economy in part concur in the importance they attach to international regimes – and we should note briefly the outlines of this convergence. Regimes have been defined by Krasner as comprising 'principles, norms, rules, and decision-making procedures around which actor expectations converge in a given issue area'.[56] While strict realists and structural Marxists tend to regard regimes as merely the instruments of the dominant actors in the system (states or classes, respectively), having no significant causal properties themselves, the protagonists of regime analysis have produced a copious literature on the preconditions for, the properties of and the conditions for stability or change in regimes.

Within this virtual sub-discipline two main strands have emerged. On the one hand, there are those who argue that such regimes are best understood as mediating variables of more basic processes and, on the other, there are those who argue that the regimes themselves may come to act as basic causal factors. In a 'modified structural' position regimes matter – in the sense that they are more than merely intervening variables – only when independent decision making (that is, uncoordinated by and through regimes) would lead to sub-optimal outcomes for the predominant actors. Depending on whether the structure of the international system is conceived in terms of global circuits of capital and the associated classes or as an anarchic interaction of states, this understanding of regimes can also take on either a Marxist or a realist form. A still stronger causal role is allotted to regimes in what Krasner terms the Grotian perspective. This strand of analysis argues that: 'elites are the practical actors in international relations. States are rarified abstractions. Elites have transnational as well as national ties. Sovereignty is a behavioral variable, not an analytic assumption . . . [These] perspectives accept regimes as a fundamental part of all patterned human interaction, including behaviour in the international system'.[57]

This Grotian perspective has clear and fatal implications for the explanatory coherence of realism and structural Marxism. For if regimes become so deeply institutionalized, if they take on such a dynamic of their own, and if they come to alter not just calculations of (relatively malleable) interests but also power resources, then they are surely basic causal variables in their own right. As Krasner note, this 'interactive' role of regimes means that realism has distinct explanatory limits.[58]

However, the realists have not remained still while this growing assault was made on their claims.

The Realist Counterattack

Neorealism and its Critics

Before we consider the specifics of Waltz's ambitious if procrustean theory it is worth noting that he adopts two important methodological tenets. In the first place, Waltz subscribes to a method of theory construction which has close affinities with the positivism of the Chicago School in economics. Thus, in the same way as Friedman once defended positive economics, Waltz argues that in theory construction the question is not whether the axioms isolated are realistic but whether they are useful. And, again in the classic positivist manner, usefulness is equated with predictive power. On this basis Waltz then defends a basically Popperian methodology of falsificationism.[59] This positivist methodology accords with the systems theory that Waltz proceeds to derive. (Positivism and objectivist accounts of social structure were, of course, the dominant methodological and theoretical tools, respectively, of the 'orthodox consensus' in postwar social theory.[60])

One further borrowing from economics by Waltz is also of interest. This concerns his attempt to develop an account of international politics via an analogy with the neoclassical (or orthodox) theory of markets. Waltz notes that the regularity of international politics cannot be accounted for by the manifest diversity of the internal constitution of states, or the variable foreign policies pursued, the routine and fatal mistakes of 'reductionist' theories.[61] Thus:

> A systems theory of international politics deals with the forces that are in play at the international, and not at the national level. This question then arises: With both system-level and unit-level forces at play, how can one construct a theory of international politics without simultaneously constructing a theory of foreign policy? . . . The answer is 'very easily'. . . . An international–political theory does not imply or require a theory of foreign policy any more than a market theory implies or requires a theory of the firm. Systems theories, whether political or economic, are theories that explain how the organization of a realm acts as a constraining and disposing force on the interacting units within it.[62]

And so in the systems theory advocated by Waltz, the structure of the system is governed by the interactions of its constituent parts and is understood in terms of a generative model of structure. This is so because, although only agents can act, they are constrained to act in certain ways by the system's structure through the processes of socialization and competition. The construction of a theory of these causes, which will 'bring off the Copernican revolution' in international relations, therefore 'requires conceiving an international system's structure and showing how it works its effects'.[63]

The theoretical gains which Waltz's neorealism seeks to provide have now been established and their similarity to conventional understandings of realism should be noted. First, theoretical success is to be understood in terms of usefulness, which in turn is defined by way of successful explanation and prediction. Second, system- and unit-level forces or causes must be clearly distinguished, or in more conventional terminology the distinct character of the international must be established. Third, the international political system must be marked off from the other international systems: the autonomy of the political much be registered. Where Waltz differs from realism is in his argument that each and all of these desiderata can be met in a systems theory of international politics, which rests neither on unsubstantiated assumptions about human nature nor on unverifiable notions of the national interest.[64]

What, then, is the structure of the international (political) system? To answer this Waltz draws on the theoretical approach known as 'systems theory', adopting a model of political systems which classifies their structures according to (1) the principle through which they are ordered, (2) the degree of functional specialization of the component units, and (3) the distribution of capabilities across the units. In this schema, domestic polities are characterized as hierarchically ordered structures of specialized units which possess varying capabilities. By contrast, the international polity is comprised of an anarchically ordered structure of functionally equivalent units which differ solely by their relative capabilities.[65] (The specification of relative capabilities – that is, the distribution of power across the units – is a feature of the system, not of the units. But an operationalization of the theory requires that power is simultaneously regarded as a possession of the units. Ashley formulates this conception very well: 'For the neorealist, the state is *ontologically prior* to the international system. The system's structure is produced by defining states as individual unities and *then* by noting the properties that emerge when several such unities are brought into mutual reference'.)[66] Such is the basic conception of international political structure with which Waltz seeks to fashion a theory of the substantive processes of world politics.

Waltz further elaborates his Copernican reformulation of international relations theory by showing the effects of structural causes in the economic and military domains. In the first place, Waltz notes that the flow of trade measured as a percentage of national product was greater in the multipolar era before the First World War than in the current bipolar order.[67] And drawing on Cooper's distinction between two kinds of interdependence – sensitivity and vulnerability – Waltz concludes that 'even though problems posed by sensitivity are bothersome, they are easier for states to deal with than the interdependence of mutually vulnerable parties'.[68] On this basis Waltz charges, with some justice, that the transnationalists have failed to distinguish properly between 'variations of interdependence within a system of low interdependence' and 'difference[s] between systems'.[69] Waltz's arguments are equally parsimonious in the military field. Thus he notes that power balancing 'is differently done in multi- and bipolar systems. . . . Where two powers contend, imbalances can be righted only be their internal efforts';[70] that a clear distinction should be drawn between 'the formation of two blocs in a multipolar world and the structural bipolarity of the present system';[71] that bipolar military competition takes a zero-sum form; and that the advent of nuclear weapons has not altered the usefulness of force (not to be confused with its usability). Finally, Waltz suggests both that powerful states are not constrained to subordinate ideology to interest and that in the postwar period the United States provided for itself and many others a public good, through its role as the world's policeman.[72]

In the debate on Waltz's work and the earlier transnationalist critique of realism a number of points recur. As we saw above, Keohane and Nye had already argued that realism is weak in accounting for change, particularly when this originates in the dynamics of the political economy, or in domestic political processes. Keohane's work on international politics, *After Hegemony*, also argued that for a strong and parsimonious version of realism to be true, then (a) power in the international system must be homogenous and fungible, and (b) states must act rationally in the pursuit of their interests.[73] Finally, the transnationalist critics noted that realism was unable to account for the power and role of non-state actors at the international level. Turning now to the debate on neorealism it is clear that these criticisms reappear as theoretical objections to Waltz's model.

Firstly, critics have noted that Waltz's rigid separation between unit- and system-level processes reproduces at a theoretical level realism's inability to account for change in the structure of the international system. In Waltz's model the only element of variation is introduced by changes over time in the distribution of relative capabilities among the component units. The system can be described as bi- or multipolar depending on the allocation of power resources among its major states. Within a system in which the structure is given, however, there is no way within the system-level theory of accounting for change. Unless they radically alter the distribution of capabilities, unit-level changes by definition are not relevant, and any attempt to connect unit- and system-level changes in a theory of foreign policy is precluded by the basic conceptual framework. And anyway this focus on changing capabilities provides no theory of how system-level changes occur. Thus Waltz resorts to an untheorized voluntarism in order to explain change in an otherwise structurally determined system. For Waltz's definition of structure implies that systemic change must ultimately be located at the level of changes in the component units: 'Through selection, structures promote the continuity of systems in form; through variation, unit-level forces contain the possibilities of systemic change'.[74] But the limits to his ability to *theorize* change are starkly constrained by the deficient objectivist conception of social structure indirectly taken over from Durkheim: 'With skill and determination structural constraints can sometimes be countered. Virtuosos transcend the limits of their instruments and break the constraints of systems that bind lesser performers'.[75]

Ruggie argues that Waltz's model can be rectified in this regard by introducing an account of the functional differentiation between units in the original model of the system's structure.[76] This allows Ruggie, unlike Waltz, to distinguish between the mediaeval and the modern states-system. However, such tinkering with the Durkheimian framework is scarcely productive of theoretical cogency. Specifically, it still leaves us with the problem of how to account for change *within* the modern states-system in general (apart from through changes in capabilities) and within the bipolar postwar order in particular. In the case of the latter all that can be accommodated is, once again, changing capacities – unless Ruggie is prepared to claim that modern states are not functionally equivalent, or perhaps that relations among them are no longer anarchic. But if either is the case, then Waltz's distinction between unit- and system-level analysis collapses. One answer to these problems is to follow Waltz and simply deny that anything significant has changed.

Secondly, because Waltz's structure is defined in terms of an exclusive ordering principle of anarchy he overrates the homogeneity and fungibility of power. Keohane has argued that for a strict realism to hold

true power must be homogenous and fungible.[77] Waltz recognizes that
the role of military power has altered but simply claims that it hasn't
altered enough to make any substantive difference. Keohane tries to
take the argument further by also considering the economic dimension,
while introducing a distinction between power as resources and power
as control over outcomes. Keohane and Nye, as we saw above, tended
to argue that the former is a good predictor of the latter under realist
conditions, but that when international regimes and organizations play
a mediating role then resources and outcomes may diverge quite significantly. This is an important element towards a critique of realism,
but in itself it does not provide an adequate conception of power. To
treat 'power as resources' and 'power as control over outcomes' as two
theoretically separable types of power is to mistake a methodological
distinction between levels of social analysis for a theoretical one. It is
not that we need several theories of power (although the resources and
relations involved in the constitution of power are *not* unitary), but
rather that the place of power in social (re-)production must be apprehended both at different levels of analysis and in different institutional
domains (see below).

Thirdly, Waltz's insistence on the functional equivalence of the component units excludes from his model any recognition of the agency of
non-state actors at the international level. For while Waltz accepts that
his model cannot explain the entirety of international politics, he insists
that aspects raised by critics properly belong to unit-level processes and
cannot, therefore, compromise his structural theory of the balance of
power. It is precisely this point which Keohane and Nye, in their
reconsideration of the arguments of *Power and Interdependence* a decade later, seek to challenge: 'Such factors as the intensity of international independence or the degree of institutionalization of international rules do not vary from one state to another on the basis of
their internal characteristics . . . and are therefore not unit-level factors
according to Waltz's . . . definition'.[78]

More generally, this point can be reformulated by noting that the
concept of the 'state' used in realism is highly problematic. Here the
Grotian school of international regimes analysis seems to have gone
farther than most in the field of international relations. In this perspective, again as noted above, the state's sovereignty is explicitly treated
as a behavioural rather than an analytic variable. A somewhat similar
recognition is involved in Keohane and Nye's perception that the state
need not be treated as a unitary actor, that it is composed of specific
institutions which may to a greater or lesser degree function independently of one another, that powerful non-state actors may come to

influence the behaviour of parts of the state apparatus, and that these
aspects cannot sensibly be consigned to unit-level, domestic processes.

Krasner's earlier work, *Defending the National Interest*, addressed a
similar set of problems.[79] In this extraordinarily rich book Krasner
argued that the use of the fundamental realist axiom – that in a
condition of anarchy, states pursue the national interest defined as
power – in a logical–deductive model of international politics was
empirically limiting, since it could not explain state behaviour when
the core objectives of territorial and political integrity were not at
stake. According to Krasner, this limitation was particularly strong for
a secure hegemon such as the United States. Instead, he suggested that
the national interest should be defined inductively, from a number of
carefully selected tests in a case study examination of foreign policy
making and execution. In this way, Krasner sought to vindicate realism
not by arbitrary fiat but by a concrete analysis of US policy in the
field of raw materials. Unlike deductive realism, therefore, he neither
conflates the nation with the state, although he does argue that the
executive branch of the state consistently pursued a policy of the
'national interest' (irreducible to that of powerful economic interests
and coherently if ideologically related to broad issues of security), nor
does he conflate the state and the government, in so far as he demonstrates the repeated struggles between the Congress and the institutions
of the national security state – principally the National Security Council,
the CIA, the Pentagon, and the State Department.

Fourthly, Waltz's obduracy in clinging to such a restrictive set of
founding assumptions can be seen as inconsistent with his own methodological tenets. The purpose of adopting the restrictive axioms of neoclassical economics, which Waltz models his own procedure on and
draws his analogies from, is both to capture the most significant
elements of the processes under study and to render analysis tractable.[80]
Progress in economics is then held to arise through relaxing the assumptions such that the predictions of the theory come closer to observable
features of the world. Through such a procedure the explanatory content of the theoretical framework is enriched. A procedure whose rules
demand that theorists endlessly restate and refine the basic axioms,
while insisting that if the world doesn't look like the model then it
damn well ought (to be compelled) so to do, is strictly for ideologues;
and it is the retreat into formalism entailed by this process which has
contributed to the increasing irrelevance of much recent economic
theory.[81] International relations will gain nothing from following this
path.

Rather, if Waltz took his adherence to a Popperian stance seriously, he would surely follow Keohane's procedure and adopt the sophisticated methodological falsificationism espoused by Lakatos: namely, a theory is only falsified when a new rival has a surplus empirical content, explains all that was explained by the old theory, and has some of its surplus content corroborated (that is, it has withstood attempts at falsification).[82] By these criteria some of the claims made for transnationalism look pretty strong in some areas, and one could even argue that realism is only a limiting case of a more general theory. The least that can be said, then, for the critics of neorealism is that they have pursued its preferred methodology in an appropriate manner.

New Agenda, New Questions

In reality, of course, much of the debate between the realists and the transnationalists stops short even of being a dialogue of the deaf because the latter have in effect been asking the wrong questions. The transnationalist critique of realism (and the development of regimes analysis) will always be self-limiting in so far as it sees its role as challenging realist axioms on empirical and local theoretical grounds. Because of this it remains trapped in the same basic conceptual universe as realism. For at this level, transnationalism *shared* the founding assumption of the realist problematic, that states subsist in a condition of enduring anarchy. It claimed merely that the ties of interdependence progressively attenuated the causal effects of anarchy on international political outcomes. By breaking with realism only in terms of the empirical description of unit-level processes and behaviour, leaving the unitary model of the state intact at the system-level, transnationalism effected a temporary displacement of the state, but it offered no alternative to the realist theorization of the state. As long as this remains the case, it is not difficult for (neo-)realists to show that nation-states, questions of security, the pertinence of military force, and state power have not disappeared.

As Fred Halliday has suggested, the problem with this whole line of debate is that 'The argument is not whether we are not "state-centric", but what we mean by the state. . . . International Relations as a whole takes as given one specific definition: what one may term the national-territorial totality'.[83] In this version the 'state', the 'nation', and the 'government' are run together in a metaphysical concept of impregnable circularity.[84] To this, Halliday counterposes an 'alternative conception of the state' which 'denotes not the social–territorial totality, but a specific set of coercive and administrative institutions'.[85] (Of course, this is where Krasner, and on occasions Keohane and Nye, get to by a practical, empirical route.) This latter version is to be preferred because it leaves unresolved a number of questions of crucial importance in international relations that are effectively pre-empted by the national–territorial definition – what institutions in fact comprise the state, how extensive is the autonomy of the state elite, and what is the precise relation between the state and the socioeconomic order?

The real force of Halliday's point is only fully apparent in the 'second agenda' he proposes for international relations theory. For what is opened up by an explicit and alternative theoretical specification of the state is a range of themes for comparative historical research: firstly, a new look at the role of organized coercion in the origins and character of the modern nation-states system would revise realism's comfortable assumptions about the identity of states and peoples; secondly, a greater recognition of the fact that state structures are shaped by their position in the global context could begin to explore the interaction of states and the international system (however the latter is conceived); and thirdly, a study of the actual content of sovereignty can be advanced through examining the sources and types of state capacities. As an agenda for future research alone Halliday's suggestions have much to commend them, but his argument also raises in a direct way a number of new conceptual points.

By reorienting the focus from whether international relations theory should or should not be state-centric towards the question of 'what we mean by the "state"', Halliday poses a theoretical problem that is potentially as productive of basic conceptual development as he shows it to be for renewing the discipline's substantive research agenda. For the enduring weakness of realism and its successors is that the existing conceptual framework has no means of posing as a theoretical question the relations between the state and the international system. By contrast, as Halliday himself notes, the adoption of a sociological conception of the state points immediately to research areas involving not only questions of what constitutes the power of the state but also a parallel interrogation of the interaction of the political and military structures of the international system with the economic structure. Moreover, by this stage, what is implicitly demanded by such a challenge is a full-scale conceptual reconstruction of the notion of an international system and its component structures.

It is here that realism's weaknesses are most apparent, yet paradoxically have remained consistently unchallenged by its critics. And this in turn requires a shift away from the comparative historical and sociological formulations which Halliday invokes and into the domain

of social theory. This is so because while the relevant developments in
the state autonomy literature, the new sociology of militarism, and the
Marxist focus on the relative autonomy of the state may converge on
the theme of the political, they are motivated by divergent theoretical
and political projects. Crudely summarized, three themes have dominated the recent literature: comparative historical studies of the goal-
directed social transformations by (relatively) autonomous state managers, and the patterning role of state structures and activities on
political and social mobilization from below; a concern with the social
forms and political implications of militarism conceived as an autonomous set of social practices; and a new emphasis on the theoretical
heritage of mainstream nineteenth century social theory, and in particular the problems posed by the absence from either liberalism or Marxism of a satisfactory understanding of the generic nature of state power.
But it is only the last of these which promises to address the underlying
conceptual hiatus in international relations theory. (And it is only such
a new basis that can simultaneously make sense of the enduring insights
of the realist tradition and forestall a realist comeback at the level of
substantive theory.) Although this point is recognized in a practical
way by the current focus on state autonomy and militarism, it can be
grasped conceptually only if the positivist methodology and the systems-
theory (Waltzian) explanatory framework of neorealism are explicitly
faced and overhauled.

An Alternative to Positivism and Systems Theory

As we have seen, the basic conceptual framework and methodological
stance of neorealism is a combination of positivism and Waltzian systems theory. And a somewhat stylized reconstruction of the predominant international relations research programme attests to the way in
which the pursuit of a positivist methodology accounts reasonably well
for the journey taken from realism to transnationalism and regimes
analysis. This programme involved designing tests for the original theory, observing the results in terms of the explanatory power of alternative sets of assumptions in controlled case studies, and then modifying
the axioms to develop a higher level of explanatory content. Despite
the obvious poverty of positivism as a set of rules governing research
programmes in the social sciences, it is clearly the case that in international relations (as in orthodox economics) nothing has replaced it.
In fact, the continued adherence to the methodological procedures
sketched above has left the critics of realism trapped within the latter's

conceptual framework. Theoretical progress has been taken to consist in the construction of ever more elaborate theories within this given conceptual field. But what is now urgently required is a fundamental alteration of our underlying procedures. For the positivist stance presumes, of course, both that the positivist model of science gives an adequate account of the natural scientific method and that – in principle at least – the natural and the social sciences are similar in logical form. (Indeed, it is only this misplaced attempt to impose the methods of the natural sciences on a recalcitrant object-domain which explains the forced conjunction of scientism and normative prescription in much realist writing.)

Both of these assumptions are fundamentally misconceived. The post-empiricist philosophy of science (focused on the debate sparked off by Kuhn's work *The Structure of Scientific Revolutions*) and the elaboration of conventionalist and realist alternatives to logical positivism established that the latter cannot provide an adequate account of the natural sciences.[86] As to the philosophy of the social sciences a series of contributions – from the early Frankfurt School critique of positivism, through the positivist dispute in German sociology and the debates over the place of hermeneutics in social theory, to the controversy around Winch's appropriation of Wittgenstein, and so on – have clearly shown that the social sciences cannot be modelled on those of the natural world.[87] The social world differs from the natural in at least four respects: first, the structures of the social world exist only by virtue of the activities they govern; second, the actors of the social world are concept-bearing agents and such beliefs as people have about themselves and the world are constitutive of their agency; third, social structures are spatially and temporally limited; and fourth, social structures have other social structures as their conditions of existence and are thus complexly interrelated.[88]

These ontological differences between the natural and the social world have important epistemological implications. Perhaps the most important of these is that, unlike the natural sciences, the social sciences have a logical relation with their subject matter – namely, concept-bearing agents and the social relations they are bound up in – which rules out any possibility of making secure predictions.[89] Rather, the social sciences are necessarily explanatory in form, and progress is measured (if, indeed, these are the appropriate terms at all)[90] not by the accumulation of successful predictions, but by the reconstruction of the basic social structures, the identification of the ways these combine to determine the predominant institutional alignments of modern societies, and the identification and explanation of the forms of social action that these in turn engender.[91] This involves a search for the social mechanisms and their associated causal powers which govern the behaviour of events, in which 'abstract theory analyses objects in terms of their constitutive structures, as parts of wider structures and in terms of their causal powers'.[92] In sharp contrast to positivist methodology, the value of an abstraction is ascertained not by its usefulness in generating predictions but by whether it is a 'rational abstraction' – that is, one which grasps the basic characteristics of a social mechanism, or a 'chaotic conception' – that is, 'those which carve up the world in an arbitrary fashion'.[93]

Turning now to the problems of systems theory, the combination of an objectivist theory of structure and a voluntarist theory of action (the connections between the two being forged by the mechanical processes of 'competition' and 'socialization'), which recent work on the state practically repudiates, must be replaced. At this level, the premisses of realism centre on two monolithic and related categories: those of sovereignty and anarchy. How does realism understand these categories? Sovereignty is defined in Weberian terms as the result of a monopoly of the means of violence within a given territory – a property which is constituted domestically such that the state is conceived as a self-sufficient and completed political order. Anarchy is the resultant structural principle of the interaction of a plurality of such units in an interstate system. While Waltz provides the clearest theoretical exposition of this mechanical conception of structure it is also clearly present in the writings of Carr, Morgenthau, Bull, Wight and others, where its apogee can be found in the notion of the balance of power.

Realism, then, starts with *national states* which are (by definition) taken to exist prior to the *international system*. The system thus gains its structural properties from the mechanical interaction of its component parts. This entails an *objectivist* understanding of *structure* as *constraint* and *socialization* in which change cannot be theorized. (This should be compared to the additive aggregates produced in economic theory.) It is this that accounts for the repeated recourse to an untheorized *voluntarism* in order to explain change. For the mechanical conception of structure is subject to a number of routinely observed problems in social theory. The most basic of these is that the properties of a component unit cannot be defined independently of the structure which in part determines its behaviour. When this is attempted, adequate explanations of system-level change and relations of power among the units prove impossible because the examination of the interaction of unit- and system-level processes has already been precluded by the definition of the unit as ontologically prior to the system.[94]

The error involved here can be traced back to Durkheim. While contemporary Durkheimian approaches, as in structural sociology and systems theory, successfully repudiate his evolutionary and functionalist perspectives, they retain the notion of social structure as involving constraint and socialization. In this type of framework, the units or agents of the system are understood to be externally limited in their modes of behaviour by the structure which their interaction generates. But this gives too limited a picture of the way in which the parts and the whole may be related in a social system. In a critical discussion of systems theory, Giddens has noted that there are three conditions of social systems in which the parts are interrelated: regulation as homeostasis ('a loop of causally interrelated elements'); self-regulation as 'a homeostatic process that is coordinated through a control apparatus'; and reflexive self-regulation in which there is a 'deliberate accomplishment of such coordination by actors in the pursuit of rationalized ends'.[95] In Waltz's neorealism the balance of power functions so as to achieve homeostasis, a thermostat controlling the temperature of international competition. What is obscured from view is the active management of an international system by its component powers, together with the mechanisms by which this is possible, and the varying roles played in these processes by differently located powers in the international structure.

Waltz's neorealism attempted to offer a concept of social structure but in providing an objectivist formulation of this it sundered social structure from the actions of its component institutions – a theory of the international system without a theory of foreign policy. Critics rightly point out that the structure of the system only derives from the actions of its component institutions (states, classes, etc.) and that the real unit-level characteristics of theses institutions are inconsistent with those suggested by the system-level deductions. Now, there is nothing in transnationalism and regimes analysis which offers a non-objectivist account of the structure of the international system. The international system may have come to be regarded as complex, in so far as it has more than one structure – typically, an economic and a cultural component are added in. But where, say, an economic component has been added, it has been conceived from a liberal or mercantilist standpoint, in terms of essentially *national* economies interacting with one another. Moreover, the critics of realism remain embedded in this atomistic, mechanical and objectivist conception of structure; on other occasions they simply dispense with any conception of the structures of the international system.

Equally, because of this failure to conceptualize the component structures of the international system, the critics make no real advance on realism's notion of power as relative capability, or as a resource possessed by national units. (The Grotian transnationalists have argued that international regimes as well as states can have power but this still doesn't provide an adequate conception of power; it is rather a claim to have identified theoretically a new *site* of power.) Power, for realists, is a property possessed by the units of the international system (that is, the states), but this conception has no space for the notion that the power of a unit depends in part on its position in the structure, not merely on the resources – economic, military and ideological – contained within it.

The main alternative to this stance has been the holistic modern world systems theory of Wallerstein and others.[96] Wallterstein's *oeuvre* is a complex and evolving body of work, but its elaboration to date provides a clear contrast with the conceptual grid of realism. For it is above all to his work that Marxists have turned for a sustained consideration of historical capitalism as an increasingly *global system*. Prior to Wallerstein's pioneering studies it was still common to account for the history of the world economy in terms of the development of its (major) nation-states. For example, Rostow treats the world economy as a simple aggregate of interacting *national* economies, and each one is regarded as tracing out a similar, unilinear pattern of development.[97] Against this analytical simplification, wherein the real interdependencies of the world economy disappear, Wallerstein, following the lead taken by Baran and Frank, seeks to theorize the transnational structures of global capitalism. Thus, what radically sets Wallerstein apart is his double break with the conventional framework of analysis found in mercantilism or realism. In the first place, Wallerstein's primary focus is on transnational commodity chains and the social classes that these produce, rather than nationally bounded 'states' and 'economies'. Secondly, Wallerstein regards the capitalist world economy as a system of interrelated parts in which the major causal forces derive from the functional imperatives of the whole. Within this structure the logic of the nation-state system and the functioning of its component units is explicated in terms of the dynamics of the global economy.

Wallerstein's understanding of the state system is correctly seen as an illicit form of economic reductionism, but perhaps more important is the simple inversion of the realist problematic effected. For if realists derive the properties of the structure from the characteristics of the component parts, Wallerstein simply reverses the procedure: the actions of the parts – a single capitalist world market, a multi-centric state system, and a hierarchy of economic spaces – are derived from the

functional needs of the whole. And as in realist thought this whole or
the capitalist world economy is conceived in a radically ahistorical
manner, for despite the attention given to cyclical and secular processes
of development Wallerstein does not capture the concrete mechanisms
of capitalist expansion and competition with alternative modes of production. In other words, a more or less mechanical account of social
structure is replaced by a holistic, organicist conception. Neither
account is able to offer a non-reductionist explanation of the interrelations between part and whole in the structures of the international
system.

Redefining the Scope of International Relations Theory

The Structures of the Global System

What is required, therefore, is a conceptual framework which enables
the interaction of structure and action to be theoretically apprehended
in the context of a theory of power, and the subsequent application of
such a scheme to the task of accounting for the structures of the global
system. The theory of structuration developed by Giddens and the
transformational model of social activity adumbrated in Bhaskar's critical naturalism are both centrally concerned with providing such a
framework. Developing Giddens's argument, Thompson has suggested
that the relations between action and structure can be understood by
distinguishing three levels of abstraction in all social analysis.[98] The
first level is that of action where human agents pursue their interests,
participating in and potentially transforming the natural and social
world. The second level is that of social institutions, or combinations
of specific social relations and material resources, which provide the
framework for action and empower some agents relative to others. The
third and final level is that of social structure, understood 'as a series
of elements and their interrelations which conjointly define the conditions for the persistence of a social formation and the limits for the
variation of its component institutions',[99] and where power refers to
the structuring of the scope of institutional variation. Thus, in the
illustration given by Thompson, 'the distribution of power [between
collectivities of agents] in a capitalist enterprise [a social institution] is
"structured by" the relation between wage-labour and capital [the
capitalist social structure]'.[100]
      This model is clear enough when applied to a capitalist economy.
How can it be applied to the nation-state system? In line with the

protocols sketched above, Lovering has recently suggested that if we are to comprehend the nation-state system as having a social structure with its own irreducible causal powers, then we need to isolate an empirical feature which can serve as the basis for a process of rational abstraction (just as Marx focused on the relation of wage-labour to capital to develop an account of capitalist social structure). Lovering argues that the most cogent candidate for such a procedure is the relationship of *sovereignty*: 'Just as the powers of individual capitals could not exist apart from the capitalist economy within which they are embedded, so the powers of individual nation-states are contingent on the existence of a system of sovereign nation-states, within which they occupy a specific structured position'.[101] The nation-state system has a structure defined by the fact that each nation-state reflexively monitors the parameters of its own and others' sovereignty. Sovereignty, based on the mutually recognized right of nation-states to the means of internal and external violence, is a universally accepted ordering principle of national and international political power relations. In other words, the founding assumption of realism, that states are condemned to exist in a field of perpetual anarchy, does not adequately capture the dual character of sovereignty. And once it is recognized that sovereignty faces both ways, as it were, then the nation-state system can be seen as reflexively monitored by its component states rather than as merely anarchic.

For, as Giddens has argued, sovereignty is not a property that states have *before* they come into contact with one another:

> the various congresses involving the European states from the seventeenth century onwards, plus the early development of diplomacy, should not be seen only as attempts to control the activities of pre-constituted states. Rather, the modes of reflexive regulation thus initiated were essential to the development of those states as territorially bounded units. . . . Both the consolidation of the sovereignty of the state and the universalism of the nation-state are brought about through the expanded range of surveillance operations permitting 'international relations' to be carried on. 'International relations' are not connections set up between pre-established states, which could maintain their sovereign power without them: they are the basis upon which the nation-state exists at all.[102]

The subsequent association of the monopolization of the means of violence with, and its chronic deployment by, the nation-state is an integral feature of the modern nation-state system. In turn, this effective monopoly of violence underpins the state's generation of an administrative moment to all forms of social power, that is the political power derived from the conjunction of its coordination of accumulated coded information and direct supervisory activities. To speak of state sovereignty, then, is to refer to the scope of this moment both domestically and internationally. For the management of transnational relations by the state requires its simultaneous monitoring of domestic locales, while this process of regulation cannot be assured without interstate agreements.

Where this conception differs most starkly from realism is in its understanding (a) that states take some of their most important characteristics from their place in the system: they do not exist (either historically or in theory) prior to the system; (b) that the position of a given state in the system is *structured* such that a full analysis must examine the relationships between the activities of states *qua* social institutions and the nation-state system *qua* social structure, rather than arbitrarily assert that the two are theoretically separable: Waltz's sharp distinction between unit- and system-level processes is explicitly designed to preclude such analysis; and (c) that power is constituted in terms of the capacities of social institutions situated in the context of determinate social structures, rather than simply in terms of the relevant resources contained within each national unit. One important consequence of this alternative formulation is that the international system only persists as such in so far as it is actively monitored or managed by the units which comprise it. This raises directly the question of the hierarchy of managerial roles, if any, and specifically the problem of theorizing hegemony.

Where realism is on stronger ground is in its correct insistence that, in the modern world, military power is in many cases a *sine qua non* for the effective defence and maintenance of sovereignty. Because of this, military power is central to the reproduction of the global system, but realism's resolute conceptual national atomism has enabled it, theoretically speaking, to do remarkably little with this enduring insight. For once sovereignty is understood in terms of the administrative moment of state power armoured above all by coercion, it becomes necessary to distinguish clearly between the military balance among nation-states and the configuration of political power in the nation-state system: the political structure of the nation-state system is neither determined by nor coextensive with the world military order. Overall, these considerations suggest that the scope for institutional variation of each nation-state is structured by its position in the structures of administrative and military power: or alternatively, state behaviour is in part a function of the place of the nation-state in the international political system and the world military order.

But the transnationalist critique was not entirely beside the point. For if social structures delimit the scope of institutional variation, and if there are a number of irreducible social structures, then it is legitimate to focus on the structuring of state behaviour by the economic structure. But contra the transnationalist argument, as Wallerstein clearly shows, the global economy cannot be understood as a simple series of nationally bounded economies connected by Keynesian aggregates and transnational corporation linkages. Where this aspect of our model differs from realism is in its clear recognition that the foreign policies of states will be shaped by their imbrication in the economic structure of the global system. In particular, the transnational reach and competitive operation of the capitalist world economy, as contrasted with the national boundedness and command character of state socialist economies, has fundamental implications for the nature of, and the differences between, the foreign policies of capitalist and state socialist states. Specifically, even if it accepted that the world is bipolar, the United States and the Soviet Union are qualitatively contrasted as Great Powers. (And any theory of international politics which cannot recognize this is no longer even pretending to speak of the real world.)[103]

More generally, an important corollary of the existence of a plurality of independent or rather analytically distinct social structures, together with their conjoint determination of the range of institutional alignment, is that concrete social phenomena are causally *overdetermined* and should therefore be analysed in terms of the *conjuncture* – that is, the precise state of overdetermination.[104] At this point, however, something of the novelty of the modern state should be noted. Both Weberian and Marxist social theory have tended to operate with a threefold division of types of sanction or power: economic, coercive-political and ideological. But all forms of organized social power – that is, power relations which endure across time and space – require in addition the mobilization of a distinct administrative moment of surveillance through the conjunction of the 'coding and retrieval of information' and the 'direct supervision of the activities of some individuals or groups by others'.[105] In so far as the modern nation-state has arrogated to itself an immense apparatus of surveillance, backed by an effective monopoly of coercion and territorially centralized, then the state apparatus and state power are *sui generis* in their origins and effects. It follows that the moment of the state cannot be reduced either to its economic, military or ideological bases or to the kinds of social forces it seeks to advance and the projects it tries to pursue. Equally, however, the infrastructural power thus generated is not something 'possessed' by the state in opposition to the collectivities and structures of civil society, but is rather a relational property in which a constantly changing dialectic of control operates.[106]

States and Social Research

Thus, the field of international relations can be identified as a set of analytically independent and irreducible but substantively overlapping and interweaving social structures which conjointly determine the scope of institutional alignments and the overall characteristics of the global system. It is these component structures of the international system – the politico-administrative structure of sovereignty, the security structure of the world military order, the capitalist structure of the world economy and the global ideological structure – that both realism and its transnationalist critics fail to apprehend and theorize, preferring to focus instead on the institutions (states, transnational corporations, regimes) whose interaction produces and is reproduced by such structures. As Giddens writes, '"systemness" does not imply complete mutual connectedness',[107] and the actual degree of connectedness between different social structures and their respective causal efficacies is a matter for detailed research. This, in turn, can be conducted only through a dialogue of empirical and theoretical work in the context of a conceptual scheme which captures 'the relation between action and structure . . . by distinguishing three levels of abstraction'[108] – action, social institutions, and social structure. And, indeed, because the state is intimately if differentially connected with all four structures, there is an obvious place for such research to commence – with the theoretical specification and comparative examination of the state, the second agenda advocated by Halliday.

Unfortunately, the recent debate on the state is in some danger of creating a false antithesis between state- and society-centred perspectives. Thus, despite the considerable theoretical gulf which separates the sociological conception of the state and the realist national-territorial definition, there is a degree of common ground between the neo-Weberian claims made for the potential autonomy of the state elite, often treated as an analytic surrogate for state behaviour, and the empirical realist claim that the decisions of the state elite can be regarded in terms of those of a unified geopolitical subject. Both of these perspectives grant a considerable degree of decisional autonomy to state managers in so far as they theorize the state as having an institutional and organizational independence from society. In the former case, the state elite is regarded as possessing an autonomous source of politico-military power which can be deployed independently of (and even in opposition to) social groups and forces located outside of the state. This mobilization of the state's generic capacities draws upon its administrative or infrastructural power, and state managers generally couch such projects in terms of the 'national interest'. In the latter case, the imperatives deriving from the state's management of its geopolitical and military positions are taken to be independent from and more pressing than any pressures it might face from its domestic economy and society.

There are also important differences which should be registered: the neo-Weberians have a more explicit focus on the differing types and sources of state capacity than the empirical realists, and they are also most immediately concerned with the power of the state in relation to its domestic society, rather than the state-to-state linkages studied by realists. Of course, neither of these substantive claims about state autonomy follow directly from either the sociological definition of the state or the recognition of the independence and irreducibility of the interests and powers of the state. (Giddens, for example, explicitly cautions against both the realist view of the state as a unified geopolitical subject and the Weberian conception of state managers as radically autonomous.) Indeed, as Halliday has pointed out, one of the principal advantages of the sociological definition of the state is that it leaves open the questions of both the precise relation of the state to the socioeconomic and the connections between the state and the international system.

These statist positions are often counterposed to the society-centred approaches which allegedly characterize Marxist theories of the state. And it is certainly true that instrumentalist and structuralist Marxist accounts of the state do seek to reduce the power of the state to that of the demands of powerful economic groups and the functional requirements of capitalist reproduction, respectively. But it is a seriously misleading, if common, perception of Marxist state theory to see it only in terms of a simple dichotomy between the complex instrumentalism of Miliband,[109] on the one hand, and the structuralism of the early Poulantzas,[110] on the other. In the German 'state derivation' debate, in the later work of Poulantzas and Therborn on the institutional materiality of the state, and in the neo-Gramscian focus on hegemony a far more complex view of the capitalist state emerged.[111] This tradition begins to move away from economistic and functionalist theories of the capitalist state towards a non-reductionist account of the complex interrelations and determinations operating between the economy and the state system.

In particular, it is argued that rather than reducing the 'political' to the 'economic', an adequate Marxist theory of the state seeks to account for the partial institutional separation of the polity and the economy in terms of the historically specific development of the capitalist mode of production. To the extent that state power is often regarded as a fetishized form of the political power of capital, some forms of this school retain a residual functionalism, but the recognition of the need for a theory of the state is clear.[112] Equally, however, in so far as the state *qua* organization must mobilize both administrative *and* allocative resources, an account of state *power* cannot be pursued very far without considering its specific social content. All states in non-segmented, non-federated, industrial societies – whether capitalist or socialist – share the generic features adumbrated by Mann and Giddens, but the character of state *power* and the organization of the state apparatus differs markedly in each case.

Hegemony

One way of carrying forward these debates is to recognize that state strategy should be analyzed in terms of the pursuit of hegemony. That is to say, stable and enduring patterns of political rule and capital accumulation depend upon a continuous attempt to win hegemony. This research programme has concentrated on domestic political economy but, as the pioneering work of Cox makes plain,[113] there is no inherent reason why it should not be extended to the international arena. Some of the threads of this discussion may be drawn together and the central theme of chapter 2 introduced by concluding with a brief discussion of the questions involved in an examination of capitalist hegemony in the global system. Domestically, hegemony consists of a continuous project of leadership which seeks to construct particular state strategies in order to advance the long-term interests of the leading sectors in an accumulation strategy while granting economic-corporate concessions to a broader social base. This task involves both the creation of a 'power bloc' which unifies – around a common political and economic programme – the most powerful forces in the social formation, and the establishment of a broader complementarity between the structures institutions and actors of civil society and the polity, on the one hand, and those of the economic order, on the other – that is, the creation of an 'historic bloc'.

The major theoretical task is to identify and explain the predominant mechanisms and strategies which link together developments in civil society, the state and the economy. However, this focus on the strategies involved in economic and political leadership should not obscure the continuous process of class (and other) struggles which permanently beset capitalist forms of domination and exploitation. This demands the development of analyses whose multiple strands offer both breadth and depth. For political and economic struggles are conducted across a wide range of sites in the social formation and the manifest events of the social world are related to underlying institutional and organizational settings, which in turn are located within broader structural principles of the social system.[114]

Extending this form of analysis to the international domain demands a recognition that states are bound up with lateral relations in addition to their domestic or vertical ones. These relations include the state's place in the world market, in the community of nation-states, in the world military order, and in international cultural forms (or the global information order). And, as is plain when the sociological definition of the state replaces that found in mainstream international relations, states routinely manage their domestic positions by drawing on international resources, while simultaneously mobilizing domestic resources for their international ambitions. But, of course, the mediation of the 'domestic' and the 'international' is not confined to channels provided by state-to-state relations: economic, political, military and ideological phenomena can all take on transnational forms because of the broader sway of internal pacification and the guaranteeing of the administrative moment of social power provided by the nation-state system. Global hegemony obtains when a state that is secure domestically is able to maintain a leadership role in each of the international structures which comprise the international system (or at least in those structures which are conjuncturally dominant). In order to secure this position the hegemon will construct a series of relations involving both the foreign and domestic policies of other states which accept its leadership, and these relations will encompass all dimensions of the social formation.

2

United States Hegemony: Oil as a Strategic Commodity

The world energy system ... seems to be a classic case of the no man's land lying between the social sciences, an area unexplored and unoccupied by any of the major theoretical disciplines.

Susan Strange, *States and Markets*[1]

Introduction

Susan Strange's comment accurately captures the predominant view of research on both the 'world energy system' and, by extension, the global oil order. The predicament she refers to does not, however, reflect a lack of attention to the energy system on the part of social scientists, and the oil industry in particular is the subject of an extensive and steadily expanding literature. Rather, the central problem with much of the existing literature is qualitative rather than quantitative: the energy system, and within it the oil sector, remain poorly understood because they have been explored and occupied by forms of economic and international relations theory which are ill-adapted to this terrain.

Not surprisingly, the major analytic frameworks found in the literature on the politics and economics of oil replicate those found in mainstream economics and international relations. Until the events of the 1970s, the predominant theoretical approach was that supplied by liberal economic theory. But as the generally unexpected events of 1973–4 seemed to throw the structure and operation of the global oil industry into turmoil, there developed a keener interest in the intersection of economics and politics in the industry. This came to a head in the international regimes school (itself soon established as the leading approach in North America), which drew on liberal economics and the transnationalist critique of realism in order to show how the behaviour of the central agents in the oil market were shaped by a set of *political* principles, norms, rules, and decision-making procedures. A related policy analysis approach, this time deriving its analytic tools from the study of bureaucratic politics and the institutionalist tradition in economic theory, paid close attention to the decision-making processes of the relevant political and economic elites.

Sharply contrasted to these analyses, but still essentially concerned with the political economy of the industry, were those accounts of a Marxist, radical and mercantilist provenance. The radicals and Marxists have stressed the changing forms of capitalist penetration of the periphery, the struggles between land-owners and capital over the rent generated in the industry, and the state monopoly capitalist forms which govern the course of accumulation in the oil sector. Meanwhile, the mercantilists (and the 'OPEC economists') have focused on the long-term changes in the oil market resulting from the rising participation of the producer-states, itself interpreted as a result of post-colonial nationalism. Another school still is inspired directly by the realist tradition in which the state's behaviour in the oil industry is held to be subordinate to the broader management of its military, security and geopolitical interests. Finally, theorists associated with 'bringing the state back in' have begun to devote their attention to events in the oil world. In this instance, the different responses of various states to the dislocations of the oil shocks, considered along with varying national energy policies, have provided a suitable topic in comparative politics from which to draw lessons for the debate on state autonomy.

Rather than provide a detailed review of these perspectives and so retread the ground of the theoretical debates sketched above in chapter 1, I want to consider some recurring themes in the current debate on the oil industry: (1) the relation of politics to economics in the global oil industry; (2) the role of the state in the international oil order; (3) the question of how to theorize hegemony, and that of the United States in particular; and (4) the role of oil as a *strategic* commodity in the organization of US hegemony. Now while a theory of the state is absent from liberal economics and is cast in reductionist terms in Marxist accounts, the models of the state adumbrated by realists and transnationalists are also wanting. The diverse problems associated with these approaches can be seen with special clarity in their inability to develop an adequate theory of hegemony. Two questions are of particular importance in this context. In the first place, the evolution of the international oil industry coincided with and was deeply formed by the transition from an era of European colonial empires and British hegemony to a postwar epoch of Cold War, independent nation-states and US hegemony. Indeed, after the Second World War, crude oil came to play a pivotal role in the maintenance of US hegemony. Secondly, however, the parameters of US hegemony were very much broader than its control over oil and all that this implied, and the general management of hegemony in turn reacted back on developments in the sector. Thus, the geopolitical economy of the postwar oil industry both formed, and was formed by, the course of US hegemony. Therefore, an examination of some of the ways in which different research agendas address themselves to these themes, and a consideration of their understanding of global hegemony in particular, should lead to a more adequate theoretical account of both the global oil industry and the character of US postwar leadership.

Economics and Politics in the Oil Industry: Liberal and Marxist Accounts

The major actors in the international oil industry are clearly the large transnational corporations, the producer-states and the governments of the consuming nations. Because of this intricate mix of economic and political institutions, a central preoccupation of industry analysts has been the precise relation between the economic and political dimensions of the oil sector. And this question has been most acutely posed in the debates around the role of the transnational corporations and the state in the oil sector. Broadly speaking, liberal economists argue that economics and politics are clearly separable – the one the domain of markets, the other the prerogative of the state – while Marxist perspectives suggest that politics and economics – state policy and capital accumulation – are intimately related, arguing either that the latter determines the former or that both are (fetishized) forms of class domination and struggle.

In liberal theory, the international expansion of the major oil companies is regarded as one of the earliest examples of company expansion through direct foreign investment. Neoclassical economics sees foreign investment as a capital flow from one economy to another, a flow which supplements the levels of savings, technology and foreign exchange in the host country. Thus, Vernon noted that by 1870 the US oil industry was selling two-thirds of its output overseas and that it had secured a strong domestic position.[2] As foreign competitors – the Dutch in Asia and the Russians in Europe – challenged their overseas position, the US companies initially sought a degree of market control, but by the turn of the century they switched strategy in an attempt to control the supply of crude. Thereafter, competition among vertically integrated companies seeking stability ensured a global spread of transnational investment, as each firm sought to match the investment spread of its rivals. But in the long run the privileged position of these major oil companies was destined to erode. For the neoclassical theory of the 'product cycle' argues that the main significance of direct foreign investment is its contribution to technology transfer from the advanced to the less developed countries and to helping the latter gain access to world markets. On this basis, Vernon suggested that there would be an 'obsolescing bargain': as the industry's technology – at each stage of the industrial cycle – became standardized and therefore widely disseminated, independent producers and state oil companies would be able to enter the market, thereby weakening the control of the majors.[3] Building on ideas such as these, recent theoretical effort has concentrated on fashioning a general theory of transnational corporation investment – variously known as the internalization, the transactional, or the eclectic theory.

As Jenkins points out, 'the central argument of this approach is that TNCs exist because of market imperfections'.[4] It is because of the absence of adequate markets in technology and marketing skills, the huge scale of many investments with its resulting oligopolies, and the presence of state intervention into the workings of the economy, that transnational corporations seek to 'internalize' market imperfections by extending the scope of their operations. This theory depends on the general neoclassical claim that all 'imperfections' are exogenous to the functioning of the market order, and that the intervention of the state is a principal cause of such imperfections. Over time, technology transfer by the TNCs, and competition among them, would erode these imperfections, with the result that the market structure in the sector would become increasingly competitive. The liberal school, therefore, argues both that the behaviour of the major oil companies can be accounted for through the categories of neoclassical orthodoxy, and that the role of power in the sector is confined to the actions of producer- and consumer-states. In the long run, the economic costs imposed on those who assert 'power' over the market will increasingly constrain them to adopt market-rational behaviour. This is so because the assumptions made about markets, and the resulting identification of all imperfections as exogenous, mean that the concept of power has no place in orthodox economic theory.

Furthermore, although liberal economists recognize that the role of the state brooks large in the oil sector, they have not developed a theoretically grounded account of its activities. When it is directly addressed, the role of the state is approached via three principal routes. In the first instance, liberal writers note that oil is an exceptional (or strategic) commodity and as a result companies often receive political and financial backing from their parent governments. The presence of such activity distorts market outcomes in obvious ways. Second, at a micro-level, a range of interventions in the market are recognized: formally these can be modelled in partial equilibrium analysis. These include industry specific forms of taxation, subsidies, and nationalization, and they have their consequent impact on prices, output and resource allocation. Typical in this regard is a concern with the impact of state intervention on the optimal depletion paths for oil finds, the estimates of field profitability, and the effects of differing licensing procedures and taxation systems. And finally, at a macro-level, the exploitation of oil has important implications for gross domestic product levels, the balance of payments and the state's finances. In this case, a variety of Keynesian, Monetarist and New Classical analyses have explored the compatibility of various policy objectives and the efficacy of different policy instruments.

Once again, Vernon's account of the development of the international oil industry illustrates some of these concerns. For example, he argues that the increasing competition in the sector did not at first lead to a reduction in prices below their monopoly levels because of state intervention of a micro- and a macroeconomic kind. This included the action of the US government to protect its domestic oil industry and the balance of payments by restricting imports during the 1950s, the output restrictions of the Texas Railroad Commission in the US domestic market which served 'to prop up the price of the licensed imports and of domestic oil production',[5] the desire of the oil-exporting states for high prices and hence greater revenues, and the wish of the consuming nations to mitigate oil's competition against indigenous coal through artificially high prices for the former. As noted, there is a strong presumption in liberal theory that the costs imposed on states by virtue of their attempts to buck the market will result in a growing dominance of markets over the commodity cartels of old. Specifically, the liberal perspective in practice coheres around a number of shared propositions concerning the economics of oil.[6] Most market-based accounts argue that the price of oil is determined by the interaction of supply and demand combined with the degree of market imperfection or monopoly. The monopolistic features of the industry are regarded as temporary phenomena which are destined to be eroded by competition among the oil companies and the producing states, so that there will be a long-run tendency for prices to fall and for cartels to fail. It is also argued that any (temporary) surplus profits or rents generated in the industry will accrue to the resource owner. Lastly, *contra* the arguments of the ecologists, it is held that no special difficulties arise from the fact that oil is a non-renewable resource.

A noted exposition of these arguments for the long-run triumph of market forces in the industry is Adelman's study *The World Petroleum Market*, which argued that *decreasing* returns to scale predominated in the oil sector and so any crude price above the (competitive) market level would be subject to a downward competitive pressure.[7] Adelman's support for this argument was data showing that in the long run, and at an industry-wide level, the oil industry is subject to decreasing returns to scale. But a tendency for the erosion of oligopoly presumes decreasing returns to scale in the long run for individual firms, not for the industry as a whole. And, as Roncaglia has noted, 'nothing in Adelman's wealth of data denies that bigger firms produce at a lower unit cost than smaller firms; Adelman only maintains that increasing quantities of oil, *in the sector as a whole*, can be produced at an increasing cost'.[8] The error of Adelman's argument, at this point, was to assume that neoclassical (specifically, Marshallian) models of competition capture the fundamental aspects of pricing and production in the oil industry. A further problem in Adelman's work also followed from his strict neoclassical precepts; the absence of any consideration of the potential role of technical change in his static framework. As in general neoclassical theory, so in Adelman's argument the level of technology is taken to be an exogenous variable, and only static comparisons between stationary states are considered. But in reality technical change is in large part an endogenous feature of modern capitalist economies which move through historical time, and 'Adelman does not provide any evaluation of the pace of technical progress'.[9] It is not surprising, therefore, that in general Adelman argued that the operations of the market, not the political actions of governments or states, were all that was important for the long-run behaviour of the industry.

And just as Adelman assumed that competitive conditions mean that politics didn't matter in the long run, so orthodox approaches in general assume that the state will ultimately be constrained to act as a 'rational' economic agent. It was the patent artificiality of these assumptions that provided the underlying impetus for the attempts by the transnationalists and the policy studies analysts to theorize the role of *politics* in the industry, as contrasted with merely describing the effects of exogenous interventions in the market. In taking this route they were to some extent retreating a path already forged by what Roncaglia has termed the managerialist perspective.[10] The economic reasoning of this school was often dependent on the broad neoclassical framework, but the managerialists paid particular attention to the historical and political constitution of the oil industry. Penrose, for example, argued that, from the 1930s to the sixties, the market control of the international oil companies derived from a combination of historical and political factors. In particular, the oligopoly which the companies maintained was not seen as the result of technologically derived increasing returns to scale and barriers to entry (as Frankel had argued),[11] but was rather held to be the outcome of the joint planning arrangements in the industry.[12] As a consequence, the primary areas of study in this perspective are the managerial decisions of company elites together with the policies of the relevant consuming and producing states.

Now, despite its historical sensitivity, the major weakness associated with this school was that little or no attempt was made to provide a general explanation of the structural trends operating in the industry: the managerial approach did not provide a theory of either the long-term economic trends in the industry or the role of the state. For instance, Penrose's account of the 1973 oil crisis paid particular attention to the conflict over the state of Israel and its position in Middle East politics, the emergence of the oil-exporting countries as independent actors, and the rising power of the industrialized states of Europe and Japan *vis-à-vis* the United States.[13] But because these developments (which were undoubtedly important) were not situated in any broader theory of the dynamics of the industry or the policies of its dominant actors, the specific role and weight accorded to these factors was left unexplicated. Thus at a theoretical level, the managerial focus inevitably privileges the causal efficacy of the institutionally specific decisions of elites over that of the broader structural features at play, to the detriment of a more balanced account. Perhaps the best that can be said for this school is that it clearly displayed the limits of orthodox economic analysis, whether the latter is considered either as a theory of the operation of the main actors in the oil market, or as a suitable framework in which to model the behaviour of the state.

A significant attempt to remedy this theoretical gap in liberal accounts is Wilson's theory of the 'petro-political cycle'.[14] Wilson argues that regimes analysis and the study of elite policy-making are correct to draw attention to the complex interconnection between economics and politics in the oil industry, but he rightly suggests that neither approach gives sufficient attention to either 'the fluctuating character of the world oil market'[15] or the precise reasons why the oil industry is so highly politicized. According to Wilson, the intersection of politics and economics arises because of the strategic importance of oil in the military and economic organization of society, the size of the oil sector in the economies of most advanced states, the place of oil as the largest single component of world trade, the role of state-owned companies in the market, and the industry's periodic instability. In the theory of the petropolitical cycle it is the last of these which is seen as the most important, for it is the periodic instability of the oil sector which effectively demands some form of political regulation. The central insight of Wilson's discussion of 'world politics and international energy markets' is that the precise character of industrial organization in the sector introduces recurrent disequilibria. Thus, he notes that the dynamics of the two major price rises of the seventies clearly demonstrated that the pattern of industrial organization – understood as 'the way that companies are organized vertically and horizontally, to their ownership, and to effective control over their operation'[16] – was an important determinant of market outcomes. (By contrast, for many economists 'the structures through which transactions flow are neutral and relatively unimportant').[17] A recognition of this, Wilson avers, enables one to construct a theory that accounts for the turbulent behaviour of the oil market in the post-OPEC decade.

While Wilson is surely correct to recommend that theorists attend more closely to the question of industrial structure, this alone seems an unduly narrow base on which to construct a general theory of the politicization of the oil sector. At a purely economic level, there are additional reasons, relating to the specific features of oil production as well as the general form of capital accumulation, why the sector is routinely unstable. Of even greater importance, however, is the *strategic* nature of oil. Wilson notes this (as did Vernon before him), only to neglect it. But the fact that oil alone, at least in the postwar period, played such a large and central role in military mobility and economic development has important implications for any analysis of the structuring of the state's activity in this sector. Because of this strategic quality, US control over the international oil order played a vital role in the constitution and maintenance of its postwar hegemony.

If neoclassical economics and the managerial–political approach only display a partial understanding of the oil industry, have radical and Marxist accounts fared any better? These accounts draw on a long tradition of analysis which suggests that power plays a central role in economic life and cannot be neatly delimited to the sphere of politics, and there packaged for speedy export to political science and sociology.

For example, some analysts argued that the TNCs derive a degree of institutional power from their privileged access to capital, technology, marketing and raw materials. In turn, this allows the transnational corporations to pursue strategies of oligopoly control and transfer pricing. Equally, it is argued that TNC penetration of peripheral economies alters the pattern of local class relations and hence the character of state policy. By encouraging the emergence of a comprador class or an 'internal' bourgeoisie, the TNCs contribute to the subordination of Third World economies and states to the priorities of the leading imperialist powers. While this work has provided a wealth of empirical detail and a valuable critique of the neoclassical orthodoxy, it confines its procedures to the analysis of institutions (transnational corporations, states), and fails to relate this adequately to the underlying structures of capital accumulation. For if it is the case that neoclassical theory abstracts the analysis of economic agents from the social institutions within which they are located, it is equally true that much radical work on transnational corporations abstracts their institutional power from the structures of the capitalist world economy and nation-state system within which they operate. This is the point made by Murray when he writes that radical critics of orthodox perspectives 'have gone too far in lodging the laws with which they are concerned in firms as institutions, rather than treating the latter as the forms through which the laws of the market are manifested'.[18]

It is to this structural level of analysis that the 'internationalization of capital approach' adumbrated by Fine and Harris, Jenkins, Palloix and others is directed.[19] This perspective identifies the global expansion of the three circuits of capital – commodity, money and productive capital – with the growth of world trade, international capital movements and transnational production, respectively. These general forms and processes of global capital accumulation must first be theorized before the strategies of the transnational corporations in any given sector can be understood. And in contrast to the neoclassical claim that transnational corporations emerge in order to internalize market imperfections, '[t]he driving force which underlies international expansion is capitalist competition'.[20] Moreover, by following its analysis through to the level of the basic patterns and rhythms of capital accumulation which delimit the scope of transnational corporation behaviour, Marxian approaches also offer a radically different model of capitalist competition to that of orthodox (and many radical) accounts. Rather than defining the level of competition in terms of the number of firms in the market (what Weeks has called 'the Quantity Theory of Competition'),[21] Clifton has shown that the surplus tradition regards the competitive mechanisms of a developed capitalist economy as dependent upon free capital mobility.[22] In turn, the transcendence of the limits to such mobility (that is, those inherent in fixed capital) is most fully achieved not in the sphere of exchange, but in the large-scale organization of production – both geographically and across product ranges – and in the centralization of control over money capital through the development of the credit system. The expansion of trans-national corporations should not, therefore, be identified with increasing monopolization and decreasing competition, but with 'the increasingly competitive nature of the capitalist system' on a global scale.[23] (This does not mean that monopoly rents are never produced, but it does direct our attention to the grounding of monopolistic positions in the patterns of production, rather than in simple control over the final market.)

As it stands, however, there are three unresolved problems inherent in any attempt to develop this approach. First, there is the difficulty posed by the continuing controversy over the labour theory of value and the 'transformation problem'. Marx's own general analysis of capitalism, as well as his specific theory of rent, presupposes the validity of his value calculus together with the associated transformation problem, and this is equally true of recent attempts to develop a Marxist account of the oil industry. As we have seen, this theory is not one which determines relative prices – unlike the problematic labour-embodied theory of value magnitudes – but rather the basis for analysing capitalism as a disequilibrating process of class struggle and capital accumulation.[24] None the less, the broad framework for the Marxian analysis of the internationalization of capital, the character of capitalist competition and the struggles over the distribution of any rent stands independently of the labour theory of value considered as a theory of relative prices in equilibrium. In so far as a static theory of price determination is required, the Sraffian model developed by Roncaglia is consistent with these broader Marxian themes.[25] In this model, the rate of growth of output, the conditions of reproduction, and the rate of profit are determined separately from the structure of relative prices. As a first approximation, it is assumed that prices are set according to the requirement that each sector in the economy earn a uniform rate of profit. But, as in the general oligopolistic case, so in the case of the 'trilateral oligopoly' found in the oil industry, there are a number of possible equilibria. The actual outcome depends to a large extent on the strategy adopted by the major agents which constitute the oligopoly: namely, the producing countries, the oil capitals and the consuming nations. Roncaglia summarizes the implications of this as follows:

In order to explain the level and the evolution of oil prices we should look not only at production costs, but also and especially at those factors affecting conditions of entry and the internal power relationships of each of the three groups (producing countries, oil companies and consuming countries), as well as at interest relations (affinity or conflict) among the component members of different groups.[26]

Second, Marxian approaches generally depend on a reductionist theory of the state: the state is often the resource owner in the oil sector, the states of the consuming nations intervene directly in the industry and some capitals are either state-owned or have a degree of state participation. In general, Marxian political economy suggests that the character and role of the state depends on the dominant form of capital accumulation. For instance, whereas Lenin associated imperialism with the dominance of monopoly capital, the internationalization of finance capital, competitive colonization and inter-imperialist wars, Fine and Harris argue that postwar imperialism was dominated by state monopoly capitalism and the internationalization of productive capital.[27] In this latter phase, the state seeks to promote the global spread of its multinational capital (subject to domestic constraints), the inter- and transnational state apparatuses take on a key role as the guardians of the conditions for capital accumulation in general, and the particular role of both is overdetermined by competition among different (national or regional) blocs of capital and the power of the labour movement.

Third, as a consequence of these limitations, many Marxist accounts (along with orthodox economic ones) underestimate the degree of openness in the course of capital accumulation, thereby minimizing the relatively autonomous role of the state in shaping the trajectory of economic development. For each moment of the circuit of capital there are a set of relevant social relations and material resources (or institutions) which help to stabilize some patterns of accumulation, rather than others. Gordon, Edwards and Reich refer to this institutional patterning of economic growth as a 'social structure of accumulation'.[28] The cycle of money capital is embodied in the monetary and credit systems together with the forms of resource and labour supply; the productive cycle is cast by the politics of production into specific factory regimes; and the cycle of commodity capital is embedded in the structure of final demand and the pattern of intercapitalist competition. At each of these stages the extent and form of state intervention often plays a key role in constituting the social structure of accumulation. A related and important general point follows from this: because of the institutional shaping of investment targeting, the competitive interests of a particular capital, sector or fraction cannot be identified outside the prevailing patterns of accumulation and state intervention.[29]

Turning now to the particular case of the oil industry, Nore has argued that Marxist writing on the role of the state in the dependent oil-exporting countries can be divided into two traditions.[30] In the first, the state is treated as a land-owner appropriating a portion of the rent, while in the second, the state is conceived 'as an entity acting as an individual capitalist: . . . the confrontation in the oil industry is no longer mainly between capitalist and landlord . . . but rather between state and private capital'.[31] The former approach followed Marx's writings on the transformation of surplus profit into ground-rent in *Capital*, Vol. 3, part 6, and it provides the basis for an explanation of the struggle for the distribution of any rent generated. Similarly, the second tradition has the merit of drawing attention to the potential conflicts between state and private producers.

Bina's study *The Economics of the Oil Crisis* provides a clear illustration of the strengths and weaknesses of Marxian approaches.[32] Bina is concerned to debunk those accounts of the seventies oil crisis which construe OPEC as the prime mover, and in order to do this he focuses on the integration of the Middle East into the international oil industry. In the period 1901–50, the major oil companies penetrated and expanded in the pre-capitalist economies of the Middle East (and elsewhere in the periphery and semi-periphery). During this phase transnational capitals directly dominated the terms of the concessions, the accompanying method of payment to the producer-states was static, and the determination of royalties was arbitrary. There followed a transitional period, from 1950 to 1970, in which the Middle East witnessed a marked expansion of capitalist forms of production and a fully capitalist global oil industry was consolidated. As a consequence, in the third period (from 1970 onward), values were produced and market prices and oil rents formed at the international level.

This account of the internationalization of capital in the oil industry is next combined with Marx's theory of rent (suitably amended for the specific nature of the industry) in order to construct an explanation of the 'oil crisis'. Here, Bina argues against both the Ricardian conception of rent understood as the distributional advantage gained by producers in areas such as non-marginal land and mines, and the neoclassical theory of resource rent which hinges on the physical distinction between exhaustible and renewable commodities. Instead, Bina notes that a theory of rent must be constructed in terms of the historically specific property relations in land, and the obstacles these pose to the free flow of capital and the equalization of profit rates by competition. Within the United States, the pattern of land ownership in combination with the rule of capture established such barriers in so far as fields were often larger than the area covered by leases, so that the full benefits of a find rarely went to the discoverer, while the resulting problems of unitization encouraged firms towards intensive exploration of a given area together with increased investment to enhance recovery rates. Accordingly, Bina suggests that:

> It is probable that the concentration of capital investment on the previously discovered oil fields, and the production of reserves by way of the extension of oil fields, led to the decline of average oil recovery in the United States. Thus . . . there were two distinct possibilities for the reorganization of production within the entire industry: (1) let the majority of oil fields in the U.S. be abandoned and excluded from production in order to preserve the old value and price structure, or (2) let the prevailing conditions in such fields . . . be generalized for the entire industry.[33]

With the aid of a detailed and valuable statistical analysis, Bina is able to show that it was the second possibility which was realized; but his explanation of this outcome is finally unclear. For while he cogently argues against a range of alternative accounts, Bina's specification of the structure of property relations in the industry is poorly developed and as a result the explanation of the resolution between the two alternatives is theoretically underdetermined. The completion of this analysis demands a further examination of the character of monopoly in the industry and an interpretation of the role of the dominant state in the oil sector, the United States.

Thus, in an important extension of this argument, Fine and Harris have argued that the pattern of its early development left the international oil industry with a double structure.[34] For while non-US companies served both domestic and international markets from overseas production, US companies served the domestic market by local production and the overseas market by domestic and overseas production. On this basis, first, the US domestic market was insulated from the rest of the world through cartel arrangements and tariffs, and second, the US industry was divided between its majors (five of the total of seven) and its independents. According to Fine and Harris, there was an essential continuity of this structure after the Second World War, albeit with increased US participation in the international industry, which in turn rested upon the political and economic hegemony of the US state. However, this double structure soon came under attack: the

US domestic industry was plagued by rising production costs, and the secure control of the majors in the world market was increasingly threatened by the activities of the (predominantly US) independents. Given this, Fine and Harris point out that through an increased oil price the industry could salvage US production and create a new form of cartelization which included both the majors *and* the independents.

For this analysis, the significance of the high degree of cartelization is that it is structurally necessary for stability in the oil market, providing the basis for 'extremely complex relations between all producers covering all aspects of the industry from extraction to sales'.[35] And in turn, on this basis two possible explanations for the price rises are offered. In the first instance, the high degree of cartelization in the oil industry suggests that conspiracy theories 'should not be discounted'.[36] And second, the desire of the US state to advance the particular interests of its multinational capital and the general interests of the US economy as a whole favoured an increase in the oil price. For this would 'improve [the US economy's] competitive position relative to its industrial rivals', and also might 'improve the US balance of payments position through the recycling of petro-dollars'.[37]

Now, as we shall see in chapters 3 and 4, there is considerable truth in this argument, but overall it operates with too narrow an understanding of US hegemony and thus misreads the principal determinants of US strategy in the Middle East. In addition, the real if limited power of the producer-states is both unaccounted for and unexplained. So, despite the highly illuminating nature of its contribution to the political economy of oil, and its ability to given an account of the power of transnational corporations in the context of the transnational structures of the capitalist world economy, as a self-sufficient research programme Marxism is vitiated by a strong tendency towards functionalist and reductionist accounts of the state. All too often, the state is portrayed either as acting to sustain the general preconditions for capitalist production both domestically and internationally, or as furthering the interests of a specific capitalist class or bloc of capital – in this case, through an increase in the price of oil, the United States promotes the interest of its own (oil) multinational capital and the overall competitive position of the US economy. The United States *has* performed these roles but it has simultaneously tried to manage a global and regional hegemony which in turn has required sustained attention to geopolitical compulsions and dynamics.

On this point, realists and their transnationalist critics agree that Marxism errs in its subordination of the state to the socioeconomic, and suggest instead that the key variable in the global political economy is the more or less autonomous agency of the state. Realism, in particular, argues that the state has a series of irreducible military, security and geopolitical interests which override any economic determinations. And to the extent that state activity can pattern economic development, through its (often unintended) role in setting the social structure of accumulation, a full account of state strategy from this perspective is of considerable interest.

The Role of the State in the International Oil Order

Beyond the approaches considered so far, the principal attempts to theorize the role of the state in the international oil order derive from (1) empirical realist premisses, (2) the analysis of complex interdependence and its associated international regimes, and (3) the recent debate on the sources of state autonomy. Perhaps the finest realist consideration of the role of the state in the oil sector is to be found in Krasner's study of raw materials investments and US foreign policy in *Defending the National Interest*.[38] Krasner's argument proceeds along two main analytical tracks. To begin with, he is concerned to demonstrate that the role of the state in this field cannot be satisfactorily explained by either a pluralist or an instrumental Marxist theory of the state. In a second and more important line of argument, Krasner uses a series of case studies to test the relative merits of a state-centric, realist approach and a structural-Marxist account.

Realist Perspectives on Oil and the State

Krasner's account of the latter is drawn primarily from the work of Poulantzas, *Political Power and Social Classes* (but also from Magdoff, O'Connor and others).[39] Now Poulantzas effectively defined the operations of the capitalist state in a functionalist way: the primary role of the capitalist state is to act as a factor of cohesion in the social formation as a whole. Krasner interprets this formulation to refer to the activities of a given state in relation to a given economy. And he argues that his case studies demonstrate that: 'In cases concerning the protection of foreign raw materials investments [including oil], American policy-makers have consistently placed general foreign policy aims above security of supply, and security of supply above more competitive markets.'[40] In other words, throughout the postwar period, the US state consistently has placed general foreign policy interests not only above those of particular powerful economic interests (contra pluralism and instrumental Marxism), but also above the defence of the general coherence of the capitalist structure of the US economy and the immediate interests of its transnational capital. A prime example of this, for Krasner, is US involvement in Vietnam which damaged the structural coherence of the economy through its consequences for investment and inflation.

Taking the work of Magdoff as a point of reference, Krasner also addresses a somewhat different interpretation of the structural Marxist argument: that the role of the *hegemonic* state is to act to secure the overall coherence of *world* capitalism. This alternative formulation is rejected on two general counts. First, Krasner suggests that in both its declared foreign policy objectives and in its actual conduct the United States has not opposed *economic nationalism* in the way it has *communism*, but that both are equally threatening to the coherence of world capitalism. Second, in those instances where vital interests have been at stake, defined as the cases where the United States has been prepared to use covert or direct force, intervention has been directed at the 'political characteristics of foreign regimes',[41] it has not aimed to restore nationalized property to US ownership and control. Again Vietnam provides an example, as does the quite different case of Chile.

On the other hand, Krasner does not simply substitute a realist account for a Marxist one. For we have seen, in chapter 1, that he notes that the predictive power of realism is weak when the core objectives of territorial and political integrity are not at stake. If, however, it could be shown that central decision-makers were able to pursue a consistently ranked set of goals, which are related to 'general societal goals', then one would be justified in speaking of the state as pursuing the national interest. Krasner suggests that this is precisely what can be shown from the historical record. But there is a twist. Although the US state pursued consistently ranked goals, the key goal was defined (ideologically) as the containment of communism; it was ideological, according to Krasner, in the sense both that the relations between means and ends were often misconstrued and that indigenous developments were routinely misinterpreted as evidence of communist expansion.

At this point, rather than see US policy as simply a catalogue of errors, Krasner draws on the distinction between 'expansionism' and 'imperialism' advanced by Schurmann in *The Logic of World Power*: expansionism is driven by economic interests whereas imperialism is motivated by the ideology of Lockean liberalism, which, following Hartz's *The Liberal Tradition*, Krasner argues has a 'totalitarian' hold over US life.[42] The transition from expansionism to imperialism, so defined, in its foreign policy was made possible by the absolute material preponderance – economic and military – that the United States possessed after the Second World War. Thus, Krasner writes: 'The American fixation on stopping communism was a product of a set of ideological precepts (Lockean liberalism) and America's extraordinarily powerful position in the international system, particularly after World War II. The precepts provided the goals of policy, and America's power made it possible to pursue these ends with little regard for economic or strategic interests'.[43]

Alongside this dominant international position, the domestic relations between the White House and the State Department, on the one hand, and the Congress, on the other, meant that the imperial presidency had a relatively free hand in the formulation and implementation of foreign policy, especially in considerations relating to the application of force. (Where Congress had an input, or where policy implementation depended on the co-operation of the private sector, foreign policy was more tightly constrained by the particular interests of US transnational corporations and the requirements of the domestic economy.)

In the case of oil, Krasner suggests that US policy in the Middle East – specifically, in Saudi Arabia during the 1940s, in Iran in 1953–4, and with respect to OPEC until the winter of 1973–4 – was primarily concerned with the stability and general pro-Western orientation of the conservative oil-producing regimes, a task rendered problematic by the United States' simultaneous support for Israel. US policy in the period leading up to the events of 1973–4 maintained this stance despite the objections of the major oil companies to price increases and nationalizations. After this episode foreign oil policy became more intricate. To begin with the economic considerations of security of supply and the price of crude oil assumed a greater importance, but also energy policy became a public and a domestic issue such that the administration had to share power with the Congress.

Now many of Krasner's case studies are convincing and his arguments against pluralist and instrumental Marxist accounts appear to be decisive, but the thrust of his overall argument is weakened by its continuing attachment to certain key realist premises. Most importantly, Krasner remains within the realist problematic to the extent that he separates the units of the international system (the states) from the structures of the system itself (and thereby subscribes to the fundamental realist postulate of anarchy); in general the only relations between the two that are considered are bilateral state-to-state linkages. In addition, he understands the power of the state basically in terms of its possession of material resources. As in the case of deductive realism, therefore, Krasner fails to consider the partial determination of state strategy by the position of the state within the component structures of the global system. In terms of the distinctions between the levels of social analysis introduced in chapter 1, Krasner has a great deal to say about the *institutions* of the international system but offers precious little on its *structures*. Thus, he discusses in considerable and illuminating detail the role of governments, the effectivity of policy instruments up to and including military force and the activities of specific multinationals, but says virtually nothing about the character of the postwar structure of nation-states, the world military order and the transnational capitalist world economy. A corollary of this is that Krasner's account and critique of the structural Marxist position is in important respects poorly specified: that is to say, it operates at the wrong level of analysis, at the institutional rather than the structural level.

For the structural–Marxist argument in relation to the United States is that as a hegemonic state it operates – alongside such international state apparatuses as the International Monetary Fund and the other leading imperialist powers – so as to secure the general coherence of world capitalism, with this general function being overdetermined by the specific defence of US transnational corporations and the internal coherence of the US economy. In other words, US foreign policy will be as much concerned to protect the *general* rules of capitalist production, distribution and exchange on a global scale, as it is to defend the *particular* interests of either the US economy or its transnational corporations. (On this account US intervention in Vietnam or even Chile is altogether more understandable.) And in order to interrogate this claim it would be necessary to give a precise account of the character of and preconditions for these general rules – and this is something which Krasner's realist assumptions prevent him from providing. However, if the conjunction of Marxian political economy and state theory can elaborate the mechanisms and strategies which connect the socioeconomic to the state, the realist critique contains a rational kernel that should be noted.

Marxism breaks with the (realist) bounded conception of sovereignty and thereby recognizes the transnational structures of capital accumulation, but it lacks a developed theory of political and military power. In this respect, the post-Althusserian focus on 'the relative autonomy of the political' has had ambiguous effects. These have been well put by Mouzelis:

> *either* one ends up with a sophisticated monism by introducing some kind of 'determination in the last instance' clause [the structural–functionalism of the early Poulantzas]; *or* one avoids monism by falling into dualism

[as in much recent neo-Gramscian work]. In this latter case the political sphere is considered as ontologically different from the economic – in the sense that whereas structural determinations operate on the economic level, agency/conjunctural considerations prevail on the level of the polity. The economy is thus held not to determine political developments directly, but merely to delineate what is possible at the level of the superstructure. What actually emerges within these set limits will then depend on the *political conjuncture* – and this leaves no more room for a theorization of specifically political structures and contradictions.[44]

And while the productive current known as the 'state-derivation' debate cogently demonstrated that an adequate Marxist theory does not explain 'politics' in terms of 'economics', but rather shows how the state is an integral part of the mode of production and that the partial institutional separation of the polity and the economy is the specific form taken by class domination in the capitalist mode of production, this tradition is no more capable of offering a coherent account of the administrative components of state power than Marxism in general.[45] It is in this context that the attention which realism pays to the generic issues of geopolitics and the logic of military security stands as a constant reminder of the latent economic or class reductionism in Marxist theories of the state.

Transnationalist Perspectives on Oil and the State

The oil sector has also proved fertile ground for the development of transnationalist theory in general and it has been seen as something of a test case for the theory of hegemonic stability in particular. Realism had argued that a state attains a position of hegemony when it is materially preponderant – economically and militarily – within the interstate system.[46] Transnationalists also contend that material preponderance is the basis of hegemony, but in addition argue for the simultaneous importance of those international regimes which are provided by the hegemon.[47] Both schools agree that the decline of the hegemon is consequent upon the rising costs of the external burdens of leadership, the internal secular tendencies towards rising consumption at the expense of productive investment, and the (uncontainable) international diffusion of technological leadership.[48] Keohane, for example, has argued that because of its material preponderance in the areas of 'raw materials, control over sources of capital, control over markets, and competitive advantages in the production of highly valued goods',[49] the United

States was able to institute a series of international regimes for co-operation after the Second World War. Partners received major benefits by 'joining American-centred regimes' in the form of a stable international monetary system (Bretton Woods, International Monetary Fund), the provision of increasingly open markets for trade (General Agreement on Tariffs and Trade) and 'access to oil at stable prices'.[50]

It is one of the merits of Keohane's work that it recognizes something of the significance of the latter, and on this point he is worth quoting at length:

> The open, nondiscriminatory monetary and trade system that the United States sought depended on growth and prosperity in other capitalist countries, which in turn dependended on readily available, reasonably priced imports of petroleum, principally from the Middle East. In a material sense, oil was at the centre of the redistributive system of American hegemony. In Saudi Arabia, and to a lesser extent in other areas of the Persian Gulf, the major U.S. oil companies benefited from special relationships between the United States and the producing countries and from the protection and support of the American government. Most Middle Eastern oil did not flow to the United States, but went to Europe and Japan at prices well below the opportunity costs of substitutes, and even below the protected American domestic price. Even though the United States never established a formal international regime for petroleum, oil was of central importance to the world political economy.[51]

Moreover, this US dominance over the international oil order was 'the result of careful strategic planning by both governmental and corporate officials, with the government often taking the lead'.[52]

In its simplest form the theory of hegemonic stability predicts that as the relative material preponderance of the hegemon is eroded, so its ability to bear the costs of providing public goods in the guise of international regimes weakens. When this happens there will either be a break up of the regime and a reversion to a (realist) self-help state of affairs, or co-operation among rational egoists may still prevail if the information providing character of regimes is recognized, thus allowing states to surmount the paradoxes of collective action. Keohane suggests that the case of oil provides some evidence to confirm this theory. For the old 'company-centred regime was destroyed through the exercise of producer's state power in a tightening market',[53] while the US loss of self-sufficiency in oil was a far greater change than any loss of trade share or monetary dominance. At first, the inability of

the declining hegemon to provide order resulted in considerable instability as the producer-states sought to impose a new form of international 'regime'. But before long 'the collapse of a hegemonic regime
. . . led to new forms of co-operation, on a bifurcated basis, within
OPEC and the IEA. The system as a whole is more discordant, but
within it limited areas of institutionalized co-operation have emerged.'[54]

A more detailed case for this perspective has been made by Bull-
Berg, who has argued that the developments prior to 1973–4 were 'in
fact the culmination of a broader producer challenge to the established
rules of the international oil game'.[55] The long decade of crisis, from
the Arab oil embargo to the price collapse in 1986, constituted an
unsuccessful attempt by the producer-states to impose a new 'interven-
tionist' regime, while the United States resisted this in favour of a
defence of 'liberalist principles'. Bull-Berg, like Keohane, does not
maintain that the international oil order can be modelled simply by
looking at the politics of the regime, for in addition three other variables
must be taken into account: (1) 'the state of the *oil market*'; (2) 'the
configuration of the *global system*'; and (3) the balance of domestic
politics (especially between the independents and the majors) in the
United States.[56] In introducing these other variables, however, Bull-
Berg draws uncritically on liberal economic and realist premises, and
this in turn prevents him from giving an adequate account of the
economic, political and military logics operating in the sector.

Similar problems beset Keohane's work. For Keohane accepts the
conventional account that the prime mover in the events of the seventies
was OPEC and he also argues that the era of US hegemony is clearly
over. Both of these claims are questionable. As I shall argue in chapter
4, a fuller account of the economic, political and military structures at
work in the international oil industry suggests that, although OPEC
played an important catalytic role, other factors were at work and that
the assertion of producer-power was only one element in a complex
causal web. Equally, it is not true that the company-centred regime
was simply overturned: a closer look at the political economy of oil
suggests that the contrast drawn between a company-centred and a
producer-state-led regime, and the claim that they now coexist, fails to
capture some of the basic characteristics of the oil sector. Finally, if
the transnational spread of US capital is considered – in its monetary,
productive and financial aspects – along with the pattern of US military
and geopolitical alliances, it is far from clear that the era of US
hegemony is over.

These limitations suggest some general weaknesses of the regimes
approach. First, in contrast to the common presumption, international

regimes are not public goods and, in consequence, the maintenance costs to the hegemon, as well as the relative benefits derived by the partners, are lower than is generally supposed. Second, US hegemony always comprised far more than material preponderance combined with the ability to order regimes, and so the relative loss of preponderance and the faltering of regimes does not provide a univocal answer to the question of whether hegemony has declined. (The failure of the transnationalist approach to register this point derives from its partial attachment to the state-centrism of realism.) Third, the regime perspective shares the general weakness of realism to the extent that it also fails to analyse the structuring of the dominant institutions of the global system by the component structures of the latter. Of particular importance here is the wholly misplaced comparisons made between British and US hegemony. Before developing these points, let us turn to the third consideration of the role of the state in the global oil order: the neo-Weberian attempt to bring the state back in.

Neo-Weberian Perspectives on Oil and the State

In chapter 1 a similarity was noted between the claims made for the autonomy of the state in mainstream international relations (and in realism in particular) and in the neo-Weberian attempt to 'bring the state back in'. This convergence can be seen clearly in the literature on oil. A representative argument, in this context, is Ikenberry's discussion of 'the irony of state strength'.[57] Ikenberry distinguishes three approaches for explaining the different responses of states to the price and supply shocks in the oil market during the seventies: societal explanations, resource-dependency accounts and a model which 'focuses on the structure of the relevant political institutions and policy instruments'.[58] He suggests that the first two explanations are of limited power, but that the state-centric approach can explain much of the variation in government policy. Ikenberry's argument rests on the claim that

the preferences and choices of government officials, particularly political and administrative officials of the executive branch, may be considered an analytic surrogate for state behaviour. The state is also a structure that fixes in place channels of access to the society and economy as well as the instruments and institutions of government. The state as *actor* and the state as *structure* are related: at moments of crisis and change, as during the oil shocks, the distinctive structure of the state itself shapes and constrains the substance of strategic policy.[59]

This bold statement is qualified to the extent that structure and strategy are mutually interdependent and that the latter can sometimes reorganize the former, thereby setting 'the analytic limits of this explanatory framework'.[60]

Now, although the leading Western states agreed a common agenda of adjustment at the Washington Conference of February 1974, with their commitments to divert to non-OPEC oil, to move away from oil to other fuels and to adopt measures of conservation, each state pursued its own distinctive strategy. Ikenberry identifies three ideal-type adjustment strategies: (1) the 'neomercantilist' strategy adopted, for example, by France, of reducing oil imports and increasing domestic energy production; (2) the German and Japanese strategy of 'competitive accelerated adjustment', which involves a combination of conservation and export promotion in order to compensate for higher import bills; and (3) the US 'defensive market response' in which the government seeks to liberate the market and allow this to organize the necessary adjustment. What determines the kind of strategy adopted, according to Ikenberry, is the particular form taken by the internal state structure, the linkages that mediate between the state and the economy, and the structures of organization within the economy and society. In turn, state strength depends on 'the ability to extract resources from society and shape private actions' combined with the capacity 'to generate independent preferences', but these are not additive properties and so a unilinear ranking of state power cannot be made.[61] As Ikenberry puts it: 'Intervention may compromise autonomy; disengagement may enhance autonomy. Such observations frustrate gross national comparisons of state capacity.'[62]

In its detailed consideration of the sources of state capacity this approach provides a useful complement to the empirical realist stance, but it is also vulnerable to similar objections. To consider the preferences of state managers as a surrogate for state behaviour is clearly to conflate the concepts of government and state. Equally, those institutions which comprise the state are imbricated in social structures that the state cannot fully control, and therefore its autonomy is always relative. And to the extent that this is so, a full consideration of the constraints on the state, which derive from these structures, is required to assess the precise degree of autonomy any given state in fact has. A final and related problem is that the range of policies considered in order to define the strategy of any given state is, theoretically speaking, more or less arbitrary. For without some general theory of the role of the state and the way in which its operations are structured it is hard to see how criteria can be drawn up so as to delimit what counts as a relevant policy and what doesn't.

These problems find a specific form in Ikenberry's account. Much that Ikenberry seeks to explain in terms of state structure is also related to the different structures of each economy and the precise form of their integration into the global market. Briefly, for example: Germany and Japan were able to pursue competitive accelerated adjustment because of the success of their export sectors; and the US adoption of a market-led response is scarcely surprising given that most of the major international energy corporations are based in the United States. In addition, the range of policies considered is to a large extent set by the policy analysis agenda, for although he notes that adjustment strategies are complex and that policy is layered, Ikenberry claims that his threefold classification does 'capture significant differences in emphasis and approach'.[63] Now this is undoubtedly true, but what is omitted may be of even greater importance in the long run: specifying variants of public policy is a necessary but by no means sufficient condition for defining state strategies. To cite only a single but crucial example: how can one assess the US response to OPEC without considering its geopolitics in the Middle East in general and its massive arms sales to the oil states in particular?

The Character of US Hegemony

Our consideration of the theories of the relationship between the political and the economic, together with the specific role of the state, in the global oil industry suggests that a number of recurrent problems have undermined the development of an adequate general account of its place and role in the modern international system. Liberal economic theory offers a comprehensive theory of the international spread of oil capital which excludes both a consideration of economic power (except through the quantitative conception of monopoly) and a theoretical treatment of the role of the state. The Marxian tradition, with its concern for the patterns of the internationalization of capital, theorizing this in the context of the historical development of the capitalist world economy, provides the missing account of the structural properties of the world economy and a theory of economic or more properly class power. However, the reductionist and/or functionalist theory of the state found in many Marxist works seriously compromises the explanatory power of this school. And to the extent that the state can shape economic development, while itself being influenced by political, military or geopolitical considerations, the structural trends of the economy are relativized by the politico-military determinants of state strategy and consequently the self-sufficient approach of Marxian political economy is called into question. This is a claim implicit in the work of the realists, the transnationalists and the neo-Weberians. But, as argued in chapter 1, none of these variants considers the geopolitics of the nation-state system or the structures and processes of the world military order in a coherent way.

The way out of this last impasse is to reformulate the traditional concept of sovereignty in two related ways. In the first place, the legal fiction of the sovereign power of the state found in realism has to be replaced by a notion of administrative aspects of state power. And second, state sovereignty should be regarded as the institutional form of a structural principle whose mutual international recognition has historically been central to the consolidation of domestic rule. Organized in this way, sovereignty is the structural principle which delimits the form taken by political power in the modern world. Now, while the maintenance of this administrative structure, both domestically and internationally, presupposes that the state is able to sustain an effective monopoly over the means of violence, it also involves the state's routine surveillance of its conditions of action as it seeks to order social interaction through a series of economic, military, political and ideological interventions. Furthermore, this combination of surveillance and a coercive monopoly, or the administrative moment of the state, is irreducible to its coercive component, and it is therefore essential to distinguish the structure of the nation-state system from that of international military power.

The world military order itself comprises three elements: the military balances, the global alliance systems, and the international diffusion of weaponry. The constitutive framework of sovereignty in the modern international system, along with the continuing industrialization of the means of warfare, do provide a channel for the diffusion of military power as a means for pursuing state strategies. And one important component of these projects is a concern for security, in so far as states undoubtedly face genuine security dilemmas which they seek to meet with internal military mobilization and external alliances. But it is a founding error of strategic studies, one to a large extent taken over from and reflecting the postwar militarization of containment, to separate the study of the complexity of state strategies from that of the logic of military competition.[64] For while there are cogent reasons for supposing that the conjunction of security dilemmas and technological progress generates a certain anticipatory, spiralling dynamic to such competition, this model is at best a partial one. The content of state strategies is overdetermined by multiple logics, and therefore it cannot be assumed that a project conducted on the terrain of military competition has as its principal dynamic a logic of security: such competition can also serve distinct political and economic ends. This is true of each component of the military order. (Contrariwise, strategies pursued in the economic domain may have as their principal motivation considerations such as those of security).

This overdetermination of composite state strategies can best be studied as a series of conjunctures, understood as the articulation of the relatively autonomous logics of the structures of the international system. A general conceptual framework for the study of international relations can license a method of study, it cannot guarantee any particular substantive theory. The search for an orrery in international relations – whether in the form of realism's balance of power, transnationalism's complex interdependence and regimes, or Marxism's laws of political economy – is fundamentally misguided. This approach is of particular interest once it is recognized that successive conjunctures mediate and represent the phased consolidation of the global system, as its component structures extend their scope and thus change their form in modern history.

These protocols suggest an alternative procedure for the study of the place of US hegemony in the modern international system. In the realist and transnationalist schools, as outlined above, hegemony is defined in terms of material preponderance combined with the ability to construct and manage international regimes, and in this context it has been widely noted that US preponderance was relatively short-lived. Russett provides the following list of declining 'power base indicators':

> loss of strategic nuclear predominance; decline in conventional military capabilities relative to the USSR, especially for intervention; diminished economic size in relative gross national product, productivity, and terms of trade with some commodity producers (principally of oil); loss of a reliable majority in the United Nations; and loss of assured scientific preeminence in the 'knowledge industries' at the 'cutting edge' and even in the numerical and financial base that enabled U.S. scholars to dominate global social science.[65]

Of course, the rapid relative decline of economic and military preponderance in the 1950s to some extent reflected the recovery of Western Europe, Japan and the Soviet Union from the enormous depredations of the war, and the relative position of the United States improves markedly if services are also counted, but there is no doubt that the distribution of resources among countries no longer privileges the United States to the extent it once did. For example, in 1950 the United States accounted for 40 per cent of world output and 27.3 per cent of world exports while in 1980 it produced 25 and 17 per cent, respectively. Similarly, the US share of world military spending fell from 51 per cent in 1960 to 28 per cent in 1980.[66] These figures need to be considered in context, however, as suggested by Huntington: 'if "hegemony" means having 40 per cent or more of world economic activity . . . American hegemony disappeared long ago. If hegemony means producing 20 to 25 per cent of the world product and twice as much as any other individual country, American hegemony looks quite secure.'[67]

Under these circumstances, liberal theorists remain confident that a high degree of interdependence can be sustained and managed through the formulation of co-operative policies within the fora of international regimes, whereas pessimistic realists predict rising sectoral protectionism, an increasing regionalization of economic activity within the world economy and a shift to (at best a benign) mercantilism.[68]

But in fact much of this argument is beside the point. For the common point of reference – a comparison of British and US hegemony – remains trapped by realist formulations. British and US material preponderance are compared and it is concluded that the former never achieved the dominance of the latter; and when the number of powers are counted it is argued that the bipolar superpower conflict is more stable than the earlier Great Power entanglements. What this approach obscures, however, is of even greater moment. The formal character of the international system had altered markedly between the era before the First World War and that after the Second World War: on the one hand, the nation-state system became truly global as the areas of formal European conquest gained independence; and on the other, the global spread of capital took on a new, genuinely transnational form as the unity of the world economy was reconstructed under US leadership. The substance of world politics was also radically altered by the Bolshevik Revolution of 1917 and the emergence of the Soviet Union as a major power after 1945. For in conjunction with continuing social and national revolts in the periphery, the systemic character of international rivalry was radically altered. If British imperialism had rivals among the other leading capitalist powers, the capitalist world itself faced weak and effectively defenceless polities on its periphery, whereas the United States faced a formidable challenge from antagonistic social forces, supported and armed to a considerable extent by the Soviet Union (and China after 1949), thereby helping to pacify intercapitalist relations. Without considering the complex impact of these profound changes, no amount of measuring and counting will begin to address the durability of US hegemony.

From European Empires to the Second World War

If hegemony has to be analysed in terms of the overdetermined position of a state within the structures of the global system, in terms of dominance conceived of as a structural directive role within the world market, the state system and the military order (rather than mere material preponderance), then we need to consider the development of these moments of the international order and the changing character of international power. As we will see in chapter 3, the origins of the international oil industry are to be found in the imperialist phase (1870–1914) of the world economy, during the period when the European powers along with the United States and Japan greatly extended their formal empires. However, the contradictions of this era, detonating both revolutions and the First World War, not only marked the ascendancy of other powers, above all the United States and the Soviet Union, but also wrought fundamental changes in the global system.[69] Prior to the imperialist epoch, the global expansion of the European capitalist powers into their periphery did not involve the formal annexation of territory 'so long as their citizens were given total freedom to do what they wanted, including extra-territorial privileges'[70] as in China and the Middle East. But with the rise of protectionism (at least outside Britain) during the Great Depression of 1873–95 and the growing concentration and centralization of capital in the core associated with the rise of finance capital, new relations began to form as these economies sought out markets and raw materials (minerals, foodstuffs and soon oil) in the periphery. And the determining process here, as Lenin and others grasped, was at once political and economic, a new conjuncture in the development of the capitalist world economy – imperialism.

Now, combined with the extraordinary material advance of Europe during the imperialist boom of 1895–1914, the 'New Imperialism' both undermined the socioeconomic stability of the periphery and destroyed its archaic polities (ancient empires, multinational autocracies and stateless orders), thereby prompting the onset of formal colonialism as well as preparing the ground for a wave of revolutionary developments of which the Russian Revolution of 1905 was the first. Recent strategic accounts of imperialism do not undermine this judgement, for as Hobsbawm explains in the British case:

> speaking globally, India was the core of British strategy, and . . . this strategy required control not only over the short sea-routes to the subcontinent (Egypt, the Middle East, the Red Sea, Persian Gulf and South Arabia) and the long sea-routes (the Cape of Good Hope and Singapore), but over the entire Indian Ocean, including crucial sectors of the African coast and its hinterland. : . . [But] India was the 'brightest jewel in the imperial crown' and the core of British global strategic thinking precisely because of her very real importance to the British economy. This was never greater than at this time, when anything up to 60 per cent of British cotton exports went to India and the Far East, to which India was the key – 40–45 per cent went to India alone – and when the international balance of payments of Britain hinged on the payments surplus which India provided.[71]

As the new stage of capitalism consolidated itself, the attendant national rivalries, now intertwined with the inevitable collapse or revolt of peripheral formations, underlay the formal colonization of 1880–1914. In turn, colonial disputes between the rival powers 'precipitated the formation of the international and eventually belligerent blocs'.[72] Meantime, in Europe the formation of the German Empire (1864–71) challenged the continental balance struck at the Congress of Vienna, while the precocious German economic advance gave it global ambitions requiring 'a global navy, and Germany therefore set out (1897) to construct a great battle-fleet'.[73] Given the conjuncture just defined, and given the economic and strategic position of Britain, this could not but challenge Britain's global position. In the context of this incipient global rivalry, what Hobsbawm has termed the 'combustible material' of the periphery provided the fuse to the First World War, which in turn provided the context for the most significant revolution of the epoch, the Bolshevik Revolution of 1917.

The war itself inflicted huge economic depredations on Europe.[74] In the case of the Central Powers, the Allied blockade, the shift to a war economy, to say nothing of the economically disastrous postwar settlement and reparations, proved lasting obstacles to economic recovery. Among the Allied Powers the picture was more mixed. For Britain and France the war imposed considerable costs through the U-boat campaign together with the impact of the war economy, and in the Soviet Union huge damage was inflicted as agricultural distribution collapsed, while the Revolution and subsequent Civil War effectively destroyed the Russian economy – not until 1927–8 did the national income reach the level of 1913. But for the United States and Japan the expansion of exports turned both into creditor nations. In the sphere of international monetary and financial relations, bilateral credit and the administration of foreign exchange replaced the gold standard. Combined with the inflationary consequences of financing the war and the reparations imposed by the Allies, this left a legacy of marked instability.

Yet, as Aldcroft has convincingly demonstrated, though the war 'distorted the economic system in several ways and also aggravated the amplitude of subsequent cyclical movements . . . it did little, if anything, to destroy the time sequence or periodicity of cyclical activity'.[75] Rather, in a context where overproduction was rising in many economies and where the global financial position remained unstable, the collapse of the US boom in 1929 together with the aggravating effects of the monetary contraction imposed by the Federal Reserve System plunged the world into a deep depression, as US (and later British) foreign lending ceased provoking deflation and further protection in Europe and elsewhere. The overall effects on an increasingly 'disarticulated world economy' have been summarized by Kindleberger as follows.

> The drying up of the international capital market, tariff discrimation, foreign exchange controls, and clearing and payments arrangements sharply reduced the proportion of multilateral payments. Trade increasingly took place bilaterally and, to the extent it continued to be multilateral, was contained within bloc lines. . . . [Under these circumstances] the 1929 depression was so wide, so deep, and so long because the international economic system was rendered unstable by British inability and U.S. unwillingness to assume responsibility for stabilizing it.[76]

Once again, economic depression, this time accompanied by even greater protectionism and state intervention, exacerbated imperialist rivalries. After the war, Britain and France had extended their colonial possessions into the mandate territories of the former German colonies and the Arab areas of the dismembered Ottoman Empire. With German policy increasingly oriented towards bilateralism in central and southeast Europe, Britain and France aimed above all to maintain their empires. But when after 1936–7 German policy swung towards rearmament, together with a drive to gain control over supplies of raw materials, it became increasingly clear that German global ambitions could not be reconciled with those of the British and the French. During 1939 a number of factors conspired to enforce action on the Allies: the timing of military spending dictated that war if necessary should be imminent; the Dominions made clear their support for the Allies in any future conflict; the United States, fearful of European developments and rising German penetration of Latin America, moved closer to the Allies; and the growing US challenge to Japanese expansion in the Pacific reduced the pressure on British and French colonies. Thus, as Overy has argued, 'Danzig was not the cause of the conflict . . . [but rather] gave the moral gloss to what was in fact a decision about when was the best time to fight for Britain and France, not for Poland'.[77]

With the outbreak of War in Europe, the failure of the Allied strategy of blockade and containment, in conjunction with the extraordinary success of the German Blitzkrieg, enabled the other Axis powers – Italy and Japan – to attack British and French interests in both the Middle East and the Far East. In the Pacific, Japan's attempt to carve out an empire in East Asia was precarious once the United States began economic warfare in the summer of 1940, restricting the export of metals and especially oil to Japan. (By the 1930s, some 80 per cent of Japanese oil imports or nearly three-quarters of total requirements came from the United States.) And, as Carr has pointed out, for Japan:

> The obvious alternative source of supply was the Dutch East Indies with its oil, rubber, tin, bauxite, manganese and nickel. In Washington fears began to mount that Japan intended to seize the area, a matter of economic as well as strategic concern to the Americans who imported 80 per cent of their rubber and tin from the Dutch East Indies.[78]

By the summer of 1941 Japan was faced with the alternative of either seizing these resources or being rendered 'powerless to resist American pressure for the abandonment of her plans to dominate east Asia'.[79] Meantime, Hitler had concluded that Britain could be left while the Soviet Union was defeated, thereby freeing Japan in the Pacific and preventing the United States from aiding the British. It was thus that Operation Barbarossa (June 1941) and the attack on Pearl Harbor (December 1941) transformed the conflict into a global struggle for hegemony.

US Hegemony

Seen in this context, the bipolarity which characterized the postwar era and simultaneously registered a qualitative break with the past should be viewed in a longer context than is common in discussions of US hegemony. What is commonly understood as the nineteenth century balance of power was, as Calleo has argued, in fact two interrelated but distinct arrangements: 'the European continental balance of 1815, struck at the Congress of Vienna and confirmed at Waterloo, and the global *Pax Britannica* of the mid-nineteenth century'.[80] But this was a highly unstable political structure, as British power soon became incapable of simultaneously preserving the continental equilibrium while maintaining and defending the Empire. European predominance was in part undermined by internecine wars on the continent, as Germany twice challenged for continental hegemony and on each occasion Britain was prepared to fight on the continent rather than lose its imperial possessions; but of even greater importance was the growing power of the Soviet Union and the United States, on the one hand, and the rise of new centres of power and conflict (and eventually nation-states) in Asia and Africa, on the other. Indeed, the long-term significance of Russian–American rivalry in the Far East around the turn of the century is hard to exaggerate. As Barraclough argued:

> the events in the Far East between 1898 and 1905 had five important consequences. First, they marked the end of the long friendship and understanding between Russia and the United States, and brought them face to face as rivals in the Pacific. Secondly, they finally established the Far East as a centre of international rivalry and conflict which . . . was for the extra-European powers, particularly the United States, in many respects more important than Europe itself. Thirdly, they saw the formation of a permanent link between European affairs and world affairs and, over a longer term, the gradual subordination of the former to the latter. Hence they implied, fourthly, that Europe was losing its primacy. And lastly, they were a turning point in the process by which the system of balance of power . . . gave way to the system of world polarity, division among a multiplicity of competing and self-balancing interests to the establishment of great self-contained, continent-wide power blocks, from which rigid iron curtains excluded all extraneous powers.[81]

Russian–American competition, therefore, began in the Pacific and spread to Europe, Southeast Asia, the Middle East and elsewhere only with 1917 and the subsequent Soviet rise to power during the Second World War. In this process, the significance of 1917 was twofold: the entry into the First World War of the United States on the side of the Allies transformed a European into a global war, ensured the temporary defeat of German designs and served notice on British imperial ambitions; while the Bolshevik Revolution radically altered the terms of social and political struggle – both domestically and internationally – throughout the globe, even if (notwithstanding Rapallo, Lausanne,

China and the Hitler–Stalin Pact) the Soviet Union played a limited directive role.[82] And whilst anti-capitalist challenges in the core were successfully, if brutally in the fascist countries, contained after the First World War, the attentuation of colonial control during the Second World War, particularly marked in Asia as a result of the Japanese occupation, proved irreparable, thereby hastening the global consolidation of the nation-state system.

The consequences for postwar reconstruction were profound. Whereas the material base for the Atlantic alliance could be readily secured with the Marshall Plan, the Truman Doctrine and the formation of NATO, US problems in the periphery were quite distinct. The United States favoured rapid decolonization not only to penetrate markets but also for geopolitical reasons. For the Americans rightly perceived that European opposition to decolonization would, if supported by the United States, drive 'the peoples of Asia and Africa over to the side of the Soviet Union'.[83] But herein lay a cruel dilemma for US policy-makers: for it was often the case that the leadership of the most successful nationalist movements were, by virtue of their attention to the social bases of peasant nationalism, communists.[84] And, *pace* Krasner, communism posed a qualitatively different challenge to US designs than nationalism alone did: the former was both more likely to align with the Soviet Union internationally and more hostile to capitalist enterprise and integration into the world market. In Asia itself the management of this dilemma was compounded by the problems associated with the occupation and reconstruction of Japan.

The politics of US hegemony therefore involved (a) the management of a complex geopolitical relationship with the Soviet Union; (b) the maintenance of capitalist unity among the advanced economies, under the banner of Cold War liberalism, through the fostering of alliances, bilateral relations and the manipulation of the internal politics of allies; and (c) an opposition to European formal colonialism and anti-capitalist social and political change in the Third World. These aspects were linked in so far as the politics and ideology of anti-communism proved a potent force for western unity on US terms, legitimated American intervention in the Third World and formed the basis for a domestic consensus around the politics of an imperial democracy. In sum, the practice of 'containment' covered everything from the management of the global and systemic confrontation between the United States and the Soviet Union with its enforced bipolarity of world politics, through to the consolidation of the nation-state system and the associated social and political upheavals in the periphery.

This highly schematic account of the development of the nation-state system, and the transition from an essentially European balance of power to a bipolar, superpower order, suggests a somewhat different interpretation of the predicament of US hegemony to that offered by Gilpin, Keohane and others. The transnationalist literature in particular focuses almost exclusively on the political relations between the United States, Western Europe and Japan, arguing that the decline of the United States in relation to its allies weakens its position *vis-à-vis* the Third World. It is indeed true that the European allies and the Japanese have become politically and economically less biddable, but this overlooks the tutelage which the United States derives from its role in the East–West and Third World conflicts. For to the extent that the capitalist core faces challenges from either the Eastern bloc or hostile forces in the Third World, and to the degree that no Western power alone remotely compares with the United States, allies have limited leverage over US strategies.[85]

Economically speaking, the basic feature of US hegemony was and remains the reconstruction of the unity of the world market, together with the asymmetrical relationships of the United States within it. This unity had been destroyed by the interwar slump and the construction of closed, state-managed economic zones during the war itself. After the war, at least outside the state socialist bloc, this unity was reconstructed. This was accomplished in a number of interrelated ways: (a) the Cold War provided an unprecedented degree of political unity among the leading capitalist states, effectively pacifying their mutual relations; (b) the occupation and subsequent reconstruction of (West) Germany and Japan, combined with the aid programmes (both economic and military) to Europe and Asia, paved the way for currency convertibility and the removal of quantitative trade restraints; (c) the decolonization of the 'Third World' and the consequent removal of the European 'political zones of *monopolistic* exploitation . . . [and] the reproduction on an enlarged scale of the economic space of oligopolistic *competition*,[86] expanded the scope of the world market; (d) the technological lead of the United States facilitated the rapid catch-up growth of the rest of the capitalist bloc; and (e) favourable demand conditions were provided by the combination of new macroeconomic policies, augmented levels of government spending and the US balance of payments deficit.[87]

So, while the formal regulation of the world market through the international hegemony of the dollar, the progressive rounds of trade liberalization, and US control over international oil were undoubtedly important, the underlying substantive component of economic hegemony lay elsewhere, in the 'spectacular increase in U.S. direct investment abroad which led to the rapid transnationalization of the industrial, commercial and financial operations of U.S. core capital'.[88] By the 1970s a parallel, competitive internationalization of West European and Japanese capital was well under way. The unity of the world market and the subsequent internationalization of capital were, as Arrighi argues, the two main *substantive* products of *formal* US hegemony. Specifying the nature of the internationalization of capital in this way as important implications for the analysis of the economic dimension of US hegemony. First, in contrast to the era analysed by Lenin or the interwar period, the dominant contradiction within the world economy underlying the events of the 1970s and eighties was neither that between *national* and *international* capital nor that between juxtaposed imperialist bourgeoisies (or national economies), but rather the combination of a general crisis of the overaccumulation of capital and uneven development (in part with respect to labour power and primary products, especially raw materials) throughout the core together with the anti-imperialist advances in the periphery.[89] Second, again in contrast to earlier periods, the form taken by the crisis of capitalist hegemony has markedly altered. As Poulantzas pointed out in *Classes in Contemporary Capitalism*:

> What is currently in crisis is not directly American hegemony, under the impact of the 'economic power' of the other metropolises . . . but rather imperialism as a whole, as a result of the world class struggles that have reached the metropolitan zone itself. . . . It follows that there is no solution to this crisis, as the European [and, one might add, the Japanese] bourgeoisies themselves are perfectly aware, by these bourgeoisies attacking the hegemony of American capital.[90]

Thus, notwithstanding the fact that the US national economy is less dynamic than its principal competitors, the global spread of capital under the continuing if weakened dominance of US capital, alongside the asymmetrical position of the US economy within the international division of labour, suggest that the erosion of formal US leadership 'has not led to the end of U.S. hegemony but simply to its transformation from formal, state-organized hegemony to an informal, market-enforced/corporately organized hegemony'.[91] Viewed from this perspective, the real crisis facing the world economy in the 1980s (and as yet far from resolved) was not so much the decline of US power, but rather the more general problems of capitalist crisis and the alteration of the balance of power between *all* states and the forces – particularly financial ones – of the world market. US hegemony was not so much in decline as increasingly unilateral in action and predatory in form.[92]

Finally, let us briefly consider the world military order. In this sphere the most obvious transformation that accompanied the emergence of US hegemony was the development of nuclear weaponry, together with the advances in rocket technology, pioneered during the war. Some have argued that the level of destructiveness of nuclear weaponry, and the related development of deterrence theory as the dominant strategic doctrine shared by the superpowers, have had a transformational impact on the use of force in the international system, such that the Clauswitzean relationship between politics and war is reversed. Others have concluded against this view, either on the grounds that 'military factors of any sort have a limited impact on international relations', or because they 'see the war-preventing aspect of nuclear weapons as offset by the dangers of war that nuclear weapons themselves introduce'.[93]

Now, while the primary function of nuclear weapons is deterrence, they also have an important secondary purpose, namely active or extended deterrence. This refers, in the US case, to the 'deterrence of Soviet *conventional* attack, by threatening U.S. *nuclear* first use',[94] or to the coercion of forces without a nuclear capability at all. In other words, nuclear weapons are called upon to serve a politico-military role. MacKenzie has captured clearly the nature of this process:

> [The global competition of the United States and the Soviet Union lies] in the struggle for influence in other regions and other countries, at most in altering each other's socio-economic or political structures. . . . [In this context] what is needed is an arsenal that translates military might into political influence. . . . There is a wish to be able credibly to threaten force, a wish to coerce, or at the very least to be able to block coercion.[95]

During the period of US strategic invulnerability, the era of 'Massive Retaliation', the pursuit of extended deterrence was more or less straightforward, but this was undermined by the growth of Soviet nuclear firepower and projection capability, combined with the improvement of weapons and their delivery systems. In the NATO arena the resulting strategy of 'flexible response' was an agreement to differ rather than a solution to US–European differences, whereas in the domain of the Third World the strategy consisted in enhanced US conventional capabilities. By the late 1960s, the development of 'mutual assured destruction' resulted in a mutual paralysis of the use of force by each superpower against the other's homeland; meanwhile it was now unclear as to how deterrence could be extended. Kissinger wrote of this situation that 'our loss of strategic superiority was a strategic revolution even if the Soviets did not achieve a superiority of their own. For that, to some extent, freed the Soviet capacity for regional intervention.'[96] None the less, the pursuit of active deterrence continued; and the logic of this led, on the one hand, to the doctrines of 'limited nuclear war' and the associated attempt to establish 'escalation dominance', and on the other, to the search for first-strike or damage limitation capabilities (perhaps through strategic defences).[97]

In addition to the US–Soviet nuclear and conventional military competition, the military components of US hegemony comprised a system of alliances centred on NATO (such as the South-East Asia Treaty Organization and the Central Treaty Organization), and a series of bilateral treaties, basing agreements, arms transfers, training missions and covert operations. Finally, US predominance within the world military order has been underwritten through a leading role in the international arms trade combined with continued technological leadership. Under these circumstances, while it is again true that the spread of military capabilities reduces the preponderance of the United States, the increased use of bilateral ties of security and dependence on arms transfers – far out-reaching the scope of similar Soviet arrangements – should be noted, as should the continuing extension of the US strategic umbrellas for Western Europe and Japan. Once more. to the extent that the West faces challenges either from the East or in the South, this military centrality may compensate for any loss of preponderance, in so far as a similar tutelage to the political and economic leadership described above can be extracted.

World Oil: Strategic Commodity for US Hegemony

Thus, the complexity of US hegemony cannot be reduced to material preponderance (however measured), not even together with the ability to order international regimes; and moreover, as the partial account rendered above illustrates, a wide range of other relationships connecting the US state to the world market, other nation-states and the world military order also need to be considered. It follows that an examination of the connections between 'oil' and 'hegemony' – that is, an elucidation of oil as a *strategic* commodity – is not exhausted by noting the US preponderance in the production and control of world oil combined with its control over the company-centred oil 'regime'. Rather, in order to consider the role played by oil in the postwar order it is necessary to situate the evolution of the oil industry in these wider structures. Accordingly, in addition to its material preponderance, the US directive role in the domain of oil derived as much from a wider structural power, which in turn delimited the scope of the institutional organization of the world oil industry.

As we will see in chapter 3, the consolidation of a unified world market, incorporating the new nation-states of the ex-colonial world, displaced the closed economic circuits of European empires and opened these regions to penetration by internationally mobile productive capital. In the case of the Middle East, this took the form of the consolidation of control over the region's oil by (predominantly US) transnational capital in the form of the majors. Moreover, the Cold War pacification of intercapitalist rivalries, in conjunction with the US leadership of a complex of military alliances, generated unrivalled political and strategic power in the Third World as a whole and especially in the Middle East, exercised through military and economic transfers. In particular, the two main oil producers in the Middle East and the region's locally dominant powers, Saudi Arabia and Iran, came to depend on the United States for their military security.

Another way in which oil came to function as a strategic commodity for the organization of US world leadership is considered in chapter 5. In this instance, Cold War pacification combined with the postwar economic strength of the US economy served as a framework for the reconstruction of Western Europe and Japan, placing these powers in a position of general strategic and economic dependence on the United States. In the case of energy, the role of the US majors in supplying low-cost oil to Western Europe (especially to West Germany) and to Japan was a factor of considerable importance to bilateral and multilateral relations, given the centrality of cheap oil to the postwar boom and for military mobility. And finally, the composite rivalry of the Cold War itself facilitated the maintenance of oil as a strategic commodity, because the various attempts by the Soviet Union to extend the integrative reach of its planned economy through the oil industry failed, thereby emphasizing the contrasted nature of the state socialist bloc in terms of its inability to compete with the capitalist world economy on the terrain of the organization of interstate economic relations.

In sum, it is the central substantive argument of the rest of this book that oil functioned as a strategic commodity because of the industry's integration into the structural components of US hegemony. And theoretically speaking, only a conjunctural model of hegemony – of the kind outlined in chapter 1 and illustrated above – which is attentive to the power relations inscribed in the broad structures of the global system, can analyze this phenomenon. On this account, the substantive consensus among realists and transnationalists, both that the era of US hegemony is over (or in rapid decline) and that OPEC has decisively

TABLE 2.1   Synoptic table of components of US postwar hegemony

Economic	*Military*	*Political*
Substantive aspects	Military balance	Global politics

1 The unity of the world market (and the spread of TNCs); and
2 The asymmetrical position of the US in the world market
	1 Nuclear balance with the Soviet Union
	2 Conventional balance with the Soviet Union
		1 *US–Soviet* rivalry
		2 *Capitalist* unity under 'Cold War liberalism':
		  (a) alliances
		  (b) bilateral relations
		  (c) manipulation of internal politics

Formal aspects	Alliance system	3 *Third World*:
		  (a) opposition to European formal colonialism
		  (b) opposition to radical anti-capitalist forces

1 International role of the dollar 2 Liberalization of world trade 3 US control of world oil	1 NATO 2 SEATO, CENTO, etc. 3 Bilateral relations – treaties, basing agreements, training missions, covert operations, etc.	
	International arms trade	
	1 Dominant position 2 Technological lead	

* I am ignoring here the powerful cultural-ideological components of US hegemony; see Russett, 'The mysterious case of vanishing hegemony' (*International Organization*, 39, 1985, pp. 209–10) for a discussion of this aspect.

broken US control over the world oil industry, can only be sustained if our analytical focus is confined to the institutional level. For, as I shall argue in chapters 4 and 6, the United States has been able to compensate for these undeniable institutional changes – above all, the declining weight of US domestic oil production and the growing direct control over production by the producer-states – by drawing upon its broader structural power, thus reconsolidating a still dominant if differently oriented degree of control over the world oil industry. Table 2.1 reminds us of the structural contours of postwar US hegemony.

3

From British to United States Leadership: Oil, Hegemony and the Middle East

Introduction

To commence our consideration of the development of the (international) oil industry, a brief examination of the basic forms of capital accumulation within this sector is in order. In general, the fundamental precondition for control over the process of capital accumulation is the combined ability to allocate money capital while securing the continued appropriation of surplus labour within the production cycle. But within the extractive sector it is commonly the case that the state has control over the natural resources in the first instance, while simultaneously lacking either the capital or the resources to win the commodity concerned. For the industry as a whole, and at a general level, it follows that there will be a dual delimitation to the path of accumulation – by the producer-states and the oil companies. The final condition for successful accumulation relates to the need for realization, the cycle of commodity capital. A focus on this moment of the circuit directs attention to demand conditions and, in particular, the role of the consuming nations. In terms of its material form, the oil industry comprises five stages: exploration, production, transportation, refining and distribution.[1] And in material terms each phase is relatively independent of the others, so the pattern of their integration can vary considerably. Indeed, one way to trace the development of the industry would be to study the changing relationships between these stages. However, although this level of material organization imposes real

TABLE 3.1   World oil production by region; (%)

	1945	1973
North America	65	21
Central and South America	16	9
Middle East	7	38
China and the Soviet Union	10	17

Source: Adapted from Fiona Venn, *Oil Diplomacy in the Twentieth Century*, Macmillan, London 1986, p. 10, figure 1.2.

constraints on (and creates opportunities for) the sector's development, it is unduly deterministic to consider these phases and their interconnections in isolation from the economic and political structures in which they are entrenched.

Accordingly, in this chapter I propose to consider the development of the industry – which as suggested in chapter 2 was coincident with the shift from European empires to US global hegemony – and the manner in which oil became a *strategic* commodity. The regional shift in world oil production during our period can be seen in table 3.1.

By adopting the term strategic commodity, I mean to indicate the complex conjunction of trends by which US control over world oil became a key resource in the overall management of its global leadership. For added to its established position as the principal source of motive power for the modern military, the growing importance of oil to postwar economic growth meant that control of the industry, alongside political dominance in the dominant producing and exporting regions (the Americas and the Middle East), provided a strong material underpinning to US hegemony. And, therefore, rather than offer a simple narrative of this process, the account to follow is structured by focusing on the attempts of the leading imperialist powers to secure control over supplies of oil, the consequent struggle both for concessions and in particular for regional hegemony in the Middle East, and the patterning of the industry in the construction of US global hegemony. This overall account draws upon and develops the analysis of the emergence of the contemporary international system sketched in chapter 2, paying particular attention to the transformations in the nature of international power resulting from both the extension of the nation-state system and the shift in the pattern of foreign investment along with the changes in the dominant forms of capital accumulation. Specifically, to this end, I attempt (1) to theorize the general aspects of the political economy of international oil; (2) to outline the historical background of the industry in the context of an expanding world economy and nation-state system; (3) to situate the postwar oil industry in the formative period when US hegemony was established after the Second World War; (4) to explain how this depended upon the shift from British to US regional hegemony in the Middle East; and (5) to show the general character of the development of the postwar industry, placing the formation of OPEC in this overall process.

The Economics of Oil

Clearly, there is no simple correspondence between the cycles of capital and the material form of the industry, rather each stage can be considered as the site for a circuit in its own right. If this is done, and then the connections examined, it is possible to map the oil sector and locate the loci of key investment decisions. During the initial *exploration* phase, corporate investment decisions focus on geographical area, timing and quantity; and there are parallel decisions taken by land-owners concerned with exploitation rights and terms of access. The salient features of *production* include (in most cases) the use of relatively simple technology and modest capital requirements, the existence of high fixed to variable cost ratios, and, in a minority of cases, high-cost technology with large capital requirements (especially in the new, post-1960s areas). In the *transportation* phase it has traditionally been less costly, as well as politically more secure, to transport crude rather than refined products, and this has resulted in the location of refining capacity (and the value-added it generates) close to the main centres of consumption. The transport phase has also been important to the extent that it has formed the locus of control for the integration of the production and refining stages. Since the 1970s, however, transport costs have fallen as a percentage of the total, as a consequence of the introduction of larger tankers and real price increases for crude. As far as the *refinery* stage is concerned, the long boom saw the growth of the oil market outpace that of the optimal refinery size, and therefore the economic barriers supporting concentration in this sphere fell. And finally, within the stage of *distribution*, the maximum price of oil was set by that of competitive energy sources and the structure of final demand.

Looking at the conjunction of the circuits of capital with the organizational phases of the industry, the oil circuits are as follows. Money (M) is exchanged for exploration commodities – labour power, means

TABLE 3.2   The cartography of oil accumulation

(i)	M – C-exp – C-find – M1 . . .	M must make decisions over location, timing and quantity of investment, landowner must decide rights and terms.
(ii)	M2 – C-prod – C-crude – M3 . . . C-find	M2 must decide rate of extraction, etc. and landowner must determine any conditions not agreed in (i), e.g. rent division (subject to (iii)).
(iii)	M4 – C-ref – C-final prod – M5 . . . C-crude	M4 must realize C-final and this will involve decisions over product mix and prices and negotiations of taxes with consumer-state governments, etc.

of production, knowledge, etc. – (C-exp) to produce an oil find (C-find) which may be exchanged against M1. The new owner of C-find exchanges M2 against production commodities (C-prod) to produce crude (C-crude) which may be exchanged for M3. Finally, the owner of C-crude exchanges M4 against refinery commodities (C-ref) to produce the final products (C-final prod) which is then exchanged for M5. If the operation is vertically integrated, the exchanges C-find – M1 and C-crude – M3 do not take place, with the same capital advancing M, M2 and M4. It is the strategic decisions concerning these latter advances that will be central to the trajectory of the oil industry (assuming for the present that the exchange C-final prod – M5 is unproblematic). Thus we have the situation depicted in table 3.2, which describes the cartography of accumulation in the oil industry.

Within the classical or surplus tradition it has long been realized that sectoral profit rates do not in general match the average rate of profit. For, other things being equal, the higher the barriers to entry into a particular sector the greater the monopoly rent appropriated. In turn, barriers to entry will depend on the size and rate of expansion of the market together with the price elasticity of the commodity concerned. Further, in those sectors which involve the exploitation of natural resources, productivity is in part determined by the natural conditions which are potentially monopolizable and non-reproducible, and to this extent cannot be generalized across the sector by competition. Variations in exploitation costs and the quality of resource won will in this case give rise to differential rent. Two complications must now be introduced. To the extent that productivity depends on the application of land capital (that is, investment to win or modify natural resources) it can be generalized across the sector. Equally, the spatial aspects of land use can be modified through investments in transportation and communication.[2]

Oligopoly within the oil sector will exist to the degree that circumstances preclude intersectoral mobility of capital, and only secondly where price competition is absent. In such a context, the rate of surplus profit (monopoly rent) generated in the oil sector may be calculated as the difference between the actual rate of profit and the general or average profit rate. And, following Roncaglia, the 'minimum long-term competitive price can be approximated by the average cost of production (including an allowance for average exploratory and developments costs) of the least productive fields necessary to meet world demand, plus the competitive profit rate'.[3] Any monopoly rents thus generated ordinarily will be allocated between the three parties centrally involved in the cycles described above: the oil companies – in the form of excess profits; the land-owners (producer-states outside the United States) as royalties and/or taxes; and the consuming nations – as indirect taxes, tariffs, etc. In economic terms, the strategies of these three actors, and within each type those of its constituent members, will be motivated by attempts to maximize their respective shares of any rent generated in the industry.

More specifically, the economics of the oil sector exhibit a number of well-analysed peculiarities. The contradictory form of capitalist accumulation, which derives from the commodification of labour power and the reproduction of class relations through the circulation of commodities, places competitive pressure on capitals to revolutionize and expand the forces of production without regard to the parameters of the market. And this means that in addition to routine disproportionalities between the branches of production on the one hand, and production and consumption on the other, regular breakdowns of proportionality occur within a given branch. In the oil industry this last form of overaccumulation has resulted in a structural tendency to overproduction, specifically determined by the interaction of two phenomena. First, more oil can always be produced from existing spare capacity and/or by exploiting new sources at less than the prevailing cost of production. Second, because of the high fixed to variable cost ratios involved in the industry there is a permanent opportunity to increase output and/or cut prices (at least in the short run), thereby augmenting cash flows and/or increasing market shares. This tendency to oversupply generates a powerful incentive to introduce oligopolistic collusion. Furthermore, due to the advantages of vertical integration in an expanding market – security of supply, increased profits, integration and control over production and marketing through dominance over the transportation and refining phases, and so forth – the industry has regularly attempted to establish control from exploration to distribution. In turn, the existence of vertical integration itself presents another barrier to entry by competitors, thereby helping to sustain the existing oligopolistic structure. Finally, the fact that a concession of nature is required for oil production to commence means that another wall against entry can be erected, as well as the power of oil capital relative to the landowners strengthened, if the allocation of concession can be organized in an oligopolistic fashion.

The Historical Background

The Rise of Standard Oil in the United States

Oil was first discovered in the United States in 1859 and had been found in the Caucasus in Russia a little earlier. In North America the emerging national market, which had been consolidated through processes of settlement, war and the building of the railroads, witnessed a large-scale merger movement from the end of the Civil War (1865) until 1914. It was during the early part of this consolidation of monopoly capitalism that the material base for the rise of the oil industry was provided by the interaction of falling costs, new technology, growing industrialization and urbanization and the proliferation of new finds. For the first twenty years or so the industry was 'marked by a multiplicity of producers, refiners, and sellers of oil [but] Standard's domination by the mid-1880s made petroleum one of the most concentrated of American industries'.[4] Indeed, by 1880 Standard Oil controlled over 90 per cent of domestic production and some 90–95 per cent of total refining capacity.[5] Shaffer has argued that the fundamental causes behind the general emergence of the oligopolistic US oil industry, and of Standard Oil in particular, were the economies of scale generated in refining, the reduction in unit costs through lower transport charges for bulk shipments (and later through pipeline ownership), and the centralization of control through the monopsonist position enjoyed by a few major refiners.[6] In general, most analysts give priority to Standard's initial control over transportation, and Williamson et al. note that 'a substantial portion of the organization's income was derived from its pipeline and transport activities'.[7] By 1911 (the year of the anti-trust action against Standard) Standard Oil's control over production of crude and its share of refining had fallen to 60–65 per cent. The rapid expansion of demand and finds after 1900, combined with the fact that Standard's greatest asset was its vertical integration, meant that 'even the smallest or regionally based companies had [to have] sufficient vertical depth to insulate them[selves] from' Standard's competition,[8] with the inevitable result that Standard's dominance began to erode.

On the demand side, the principal, world-wide need for oil in the nineteenth century was as a cheap illuminant, but by the turn of the century by-products accounted for at least half of the industry's revenue. For example, the cost of oil was four to twelve times that of coal in Britain in 1900 but prices fell rapidly as new supplies were found and the techniques of refining and distribution improved. The conversion of ships to oil – simultaneous in commercial and military fleets – was also underway: 'Britain began in 1903 with vessels operating in waters near sources of petroleum – the Far East particularly – but within a decade built a world-wide storage network that permitted the use of liquid fuel throughout the fleet.'[9] Automobiles were still luxuries in Europe before the war. In fact, it was only in the 1920s that the demand for oil really took off, as the chemical industry as well as the uses of the internal combustion engine in cars and aircraft expanded. The result was that crude oil output in 1930 (1411 million barrels) was more than four times that of 1910 (327 million barrels).

The resulting industrial structure of the oil industry was profoundly oligopolistic and was in no way destroyed by the anti-trust action against Standard. Indeed, many analysts argue that the structure of the industry thus consolidated persisted into the industry after the Second World War. In this respect the conclusion of Williamson et al. is worth quoting at length:

> By 1919, *a basic pattern of industry structure had emerged*, which in the succeeding years would be filled in intensively, as the various Standard firms [produced by the divestiture] pressed for more vertical depth and began to compete among themselves for greater shares of the business. *These moves*, coupled with shifts in consumer demand for refined petroleum products and a second merger wave in the late 1920s, *were all moulded and adapted to the basic outline of industry structure that existed at the end of 1911*. (my emphases)[10]

Here, as in almost all orthodox accounts, the basic structure of the oil industry is held to result from the economic and technological character of oil production, along with its early patterning by the rise of Standard Oil. Moreover, this kind of understanding of the industry has resulted in regular, confident and routinely falsified predictions from market-oriented economists that the oil industry is becoming increasingly competitive.

Early International Expansion

Meanwhile, in the development of the capitalist world economy, the era of free trade and *pax Britannica* gave way to the first imperialist age of 1870–1914, and accompanying this imperial expansion was the first major phase of development of the mineral industry.[11] Of course, there were also direct continuities with the free-trade era: for centre–periphery exchanges remained characterized by the reciprocal flows of manufactures for primary products. In the 1890s an Atlantic circuit of money capital began to develop and this was associated with direct (as opposed to portfolio) foreign investment.[12] The pattern of domestic expansion had taken the form of merger activity across diverse geographical areas and this experience shaped the initial foreign expansion of US capital. Direct investment circumvented trade barriers and also favoured large firms, so the characteristics shaped by the domestic market could now benefit US capital in its competitive, international expansion.[13] By 1914 over two-thirds of US overseas investment was located in four sectors – mining, manufacturing, agriculture and petroleum.

New finds were made, and new entrants were able to join the industry by the ownership of crude rather than refining capacity. Similarly, overseas companies controlling their own crude were another threat – this time to Standard's dominance of foreign markets. At the close of the nineteenth century the two new oil-producing areas were Russian and East Asia: the former was controlled by Nobel and French Rothschild interests and the latter (Indonesian oil) by Dutch and English companies. North American exports, which in 1873 constituted about three-quarters of refinery output, were progressively challenged by foreign competition and rising domestic demand. In the last quarter of the nineteenth century the Asian and Chinese markets had become increasingly important for US oil, as it tried to replace its European losses after the rise of Russian oil. Initially Standard adopted a downstream, market-oriented investment profile in its international operations – an extension of the successful domestic strategy of control through the centralization and concentration of refining capacity. As a consequence of this strategy and the new geographically dispersed competition, 'Standard Oil's share of overseas crude production was less than 1 per cent. As far as products were concerned, Standard's share of the overseas petroleum market never reached one-third in the years before World War I'.[14] This pre-war expansion of the US oil industry was to a large extent simply one part of the broader process of direct foreign investment of US capital; it was not until the First World War itself that the strategic significance of oil was fully appreciated.

None the less, among the concerned imperialist powers a significant and at times vigorous oil diplomacy existed prior to the war: Britain and the United States clashed in Mexico; the British and the Russians contested in the Middle East; the Americans and the Dutch had a dispute in the Netherlands East Indies; Britain and the United States clashed again over Burma; and the British, the Germans, the Turkish, the French, the Americans, the Italians and the Iraqis struggled for influence in Mesopotamia.[15] This international competition for oil concessions during the age of empire had both a naval and an imperialist dynamic, and the resulting 'oil diplomacy' played a significant role during the First World War. Lord Curzon's comment that 'the Allies floated to victory on a wave of oil' may have been an overstatement, but security of oil supply was a major concern of the Central Powers in particular. Indeed, German designs on Romania and the Ukraine were in part motivated by the need to secure oil, and in the Treaty of Brest–Litovsk the Bolsheviks agreed to supply Baku oil to Germany while the Peace of Bucharest gave the Central Powers control over Romania's oil wells. At this time, most of the world's oil production was located in the Western Hemisphere, and this meant that distribution would be the Allies' main problem. For Britain, however, the strategic conduct of the war was shaped by her interests in the Middle East, especially after the Ottoman Empire joined the Central Powers (November 1914), with access to India and 'the long-term issue of post-war control of oil resources'[16] proving the primary concerns.

The military implications of oil were soon seen by Churchill. In 1914 he was instrumental in securing the changeover of the British Navy from coal to an oil-burning fleet and the acquisition of a 51 per cent share in Anglo-Persian (later BP) by the British government. Wartime shortages produced a similar concern over security of supply in the United States where the administration preferred to rely on the operations of the privately owned industry. However, during the war the National Petroleum War Service Committee allocated oil between dom-

estic and military uses, while the US Fuel Administration was created
in 1918 to supervise the industry. The lack of interest in direct state
intervention reflected pre-existing close State Department–oil industry
connections (alongside divisions of interest among both the integrated
and independent producers and the domestically and internationally
oriented companies), the power of the United States as the predominant
supplier of capital to the world (especially Latin America) and the
absence of a hegemonic role for the United States in the international
system. By the end of the war, the military and economic significance
of oil was apparent to all the major powers: the Dutch had control of the
Netherlands East Indies; the French increased their domestic refinery
capacity and looked for deals in Romania and Mesopotamia; and at
the San Remo conference (April 1920) Britain was awarded the mandates for Palestine and Mesopotamia, while France gained those for
Syria and the Lebanon (a deal was also struck on oil). Indeed, in the
Middle East after 1918, only in Persia and the lands of Ibn Saud did
indigenous governments have any significant control over oil concessions – and in the former, Anglo-Persian had already acquired a
position of dominance.

In the dominant oil producer (and consumer) – the United States –
the immediate postwar years were characterized by fears of an oil
shortage, and the government encouraged a policy of 'exploitation
abroad, conservation at home'.[17] During the interwar years US administrations adopted various policies to expand control over world oil – all
examples of Wilsonian, liberal capitalist free-trade imperialism. In a
dispute with Mexico the United States formulated the doctrine of
minimum duty; in the Middle East an open door policy was proclaimed;
and in the dispute with the Dutch over the Netherlands East Indies
the notion of reciprocity was invoked to gain access.[18] This liberalism
was not confined to the oil industry and it reflected the ambitions of a
rising hegemonic power whose competitiveness would ensure complete
success in open markets. As Carr has noted:

> The search for markets to dispose of surplus production remained a
> dominant theme of American commercial policy in the 1920s and 1930s
> . . . [and] America's commitment to the preservation of world peace . . .
> reflected concern about the negative effects of war on America's world
> trade as well as anxiety about the threat civil disturbances in Europe and
> Asia posed to the security of America's vast overseas investments both
> public and private.[19]

In locational terms, Latin America was the key area for oil production
and investment by US companies (increasingly Venezuela rather than

Mexico, after the dispute with the United States and revolution in the latter – by 1939 Venezuela was the largest oil exporter in the world). But in the 1920s, the Anglo-American competition centred on North Persia, Mesopotamia and Palestine, as the United States challenged British dominance in the Middle East and Britain resigned herself to effective exclusion from Latin America. In this contest, parent governments were quick to intervene on behalf of their national capitals. Thus, under the stimulus of the war, states embarked upon a fierce competition for oil concessions, especially in the Middle East which, although then a small producer, was widely recognized as a major region of reserves. However, a direct clash between the United States and Britain was avoided by the American pursuit of 'an "informal entente" with the British based on public support for private arrangements between U.S. and British corporations'.[20]

Elsewhere, the major area of production was the Soviet Union, producing some 15 per cent of world oil. Despite oil company support for counter-revolutionary movements, Allied intervention and British occupation of the Caucasus and a boycott of Soviet oil by the majors, 'by 1939 the USSR was second in importance to the United States as a supplier of oil to West Europe'.[21] Finally, conditions of glut soon replaced fears of scarcity in the United States, as a result of new and expanding production in Texas, and globally as production rose in the Middle East and Eastern Europe. In these conditions, overproduction once again threatened the stability of the industry.

The Red Line and Achnacarry Agreements

Most importantly of all, the interwar era witnessed a series of transnational agreements and domestic regulatory interventions that secured control over the non-communist world's oil by the 'Seven Sisters'. The latter comprised those companies created by the breakup of Standard Oil – Standard Oil of New Jersey (Exxon), Standard Oil of New York (Mobil) and Standard Oil of California (Socal) – together with two other US companies (Gulf and Texaco), and two more from Europe (Royal Dutch/Shell and British Petroleum). Gulf and Texaco expanded overseas alongside Standard from a base in North America, and in their drive abroad the US majors were attracted both by the large profits to be had from low-cost foreign oil and by the fact that leasing costs in the United States (usually a one-eighth royalty) were generally higher than the cost of concessions overseas.[22] To this end, US firms partially relocated investment from Latin America to the Netherlands

East Indies and the Middle East: in consequence, in the period 1919–35 foreign oil investment began for the first time to rise faster than total foreign investment.

Despite their initial lack of access to foreign crude, US enterprises were in a strong bargaining position given that they provided some 60 per cent of foreign demand for Anglo-Dutch controlled crude. In the United States, meanwhile, the tendency to oversupply was aggravated by the new finds (especially in East Texas) and the fall in demand due to the Depression. Prices fell rapidly and the situation was chaotic until the authorities (in the form of the Texas Railroad Commission) intervened and introduced 'prorationing', whereby demand was shared between producers.[23] This was crucial because without regulation of the dominant US domestic oil market there could be no international stabilization. (A Federal Oil Conservation Board was established in 1924, composed of the Secretaries of State for the Interior, Commerce, War and Navy.) In the international arena, the so-called Red Line Agreement of July 1928 involved the renewal of a holding company – the Turkish Petroleum Company, later the Iraq Petroleum Company – by BP, Shell, Exxon, Mobil, CFP and Gulbenkian. This was established to operate collusively in the area of the former Ottoman Empire. Later the agreement was extended to Abu Dhabi (1939, the ADPC), to Qatar (1935, QPC) and other places; in addition BP and Gulf had a consortium for Kuwait (the KOC), Socal and Texaco (after the Second World War with Exxon and Mobil) had a similar arrangement for Saudi Arabia (Aramco), and finally BP had the concessions in Iran until the coup of 1953.[24]

The purpose of this horizontal integration of upstream activity was to match supply to demand, and to control the allocation of concessions, thereby deriving a high degree of control over the pattern of accumulation in the industry for the majors. This arrangement was considerably strengthened by the nature of the concessions won by the consortia: for the concessions generally granted complete control over production across extensive areas for 60–90 years, while giving oil capital control over pricing with exemption from taxes and duties. In return the landowners received a fixed royalty of 20–25 cents per barrel produced.[25] As Al-Chalabi notes, an important consequence of these arrangements was that the 'inter-territoriality [which] governed inter-company investment behaviour [meant that the] host countries were left to act merely as tax collectors in their individual territories'.[26] In total, the conjunction of US domestic prorationing with the Red Line Agreement also meant that the companies could plan Middle East, and indeed global production, to meet projected demand.

To complement the majors' control of production a secret marketing agreement was also concluded in 1928, the 'As Is' or Achnacarry Agreement. Initially involving only Shell, BP and Exxon, this was soon extended to the other Seven Sisters and the Compagnie Française des Petroles (CFP), and it was to last until the early 1950s when it was exposed due to the enquiries of the US authorities. This incestuous arrangement centred around '(a) fixing quotas; (b) making adjustments for under- and overtrading; (c) fixing prices and other conditions of sale; and (d) dealing with outsiders'.[27] Together with the Red Line Agreement and US domestic prorationing, the Achnacarry Agreement generated large surplus profits in the oil sector. For the price of crude in each market was determined by one criterion, wherever the origin of that crude, namely the prices of crude from Texas points of export in the Gulf of Mexico plus the transport costs to the country of destination.[28] And since the cost of production in the Middle East was only a fraction of Texas costs large monopoly rents were generated. Through these agreements, then, the industry was rapidly dominated by the majors and as late as 1972 they controlled 91 per cent of Middle East output and 77 per cent of the non-communist world's oil supply outside the United States.[29]

Meanwhile, the structure of the US domestic industry developed somewhat differently, producing a trichotomy of the top eight, the lesser majors and the non-integrated independents.[30] Superficially, the US domestic oil industry exhibited only a moderate level of concentration; however, in fact it was integrated by a variety of joint ventures (such as joint ownership of pipelines) and intercorporate interlocks. Further, in addition to the short-term advantages of prorationing – namely, higher prices and a reduced pressure for efficiency – domestic oil had the benefit of some of the most advantageous tax provisions ever devised. The combination of percentage depletion and the ability to base tax deductions not on costs but on selling prices allowed the recovery of ten to twenty times the cost of a field over its lifetime; the expensing of intangible drilling costs resulted in further considerable gains; and finally, the foreign tax credits provided large benefits in the period after the Second World War.[31]

Alongside the absolute dominance of US oil within world production, the control engineered under the Red Line and Achnacarry agreements was only possible on the basis of the power of metropolitan capital within the imperialist phase of the global economy, together with the as yet partial consolidation of the nation-state system in the periphery. For the pattern of these arrangements and concessions was not the result of competition between, or simply collusion among, the oil

companies, but rather involved the direct intervention of states acting
on behalf of their respective capitals. In this interstate competition the
struggle between the rising hegemonic power – the United States – and
the older European imperial powers was of central importance. Indeed,
as we shall see, these transnational regimes presupposed the temporary
dominance of British and French interests over the former Arab territories of the Ottoman Empire – passed on under the League's system
of mandates – in combination with a period of relative equilibrium in
the competitive process whereby the United States served notice on
the closed trading arrangements of the British Empire. Within this
order, the oil industry – both domestically and internationally – was
the privileged recipient of direct and indirect state intervention: oil
capital was, in other words, embodied in state monopoly circuits of
capital accumulation.[32]

Industry Structure on the Eve of the Second World War

These developments resulted in what Fine and Harris have termed a
'double duality' in the structure of the global oil industry on the
outbreak of the Second World War: the division between the US market
and the rest of the world, and the separation of domestic US producers
and the US majors. These divisions were not coincident because the
'US majors served both foreign and domestic markets through foreign
and domestic production respectively'.[33] Domestic producers lobbied
for an oil tariff from 1929 to its passing in 1932, while the US majors
could not improve their position by increasing domestic production,
and so they required Middle East reserves to supply their overseas
operations. Indeed, domestic regulation, through the Interstate Oil and
Gas Commission and the Conolly Act (both were passed by Congress
in 1935, and the Supreme Court recognized the constitutional legality
of prorationing), limited US production and therefore allowed the
insertion of the domestic industry into the global industry.[34] Now,
although the US share of Middle East oil was only 16 per cent at the
outbreak of the Second World War, this must be considered in the
context of the fact that in 1940 US domestic production constituted 73
per cent of non-communist output, whereas Middle East production
was a mere 6 per cent.[35] Equally, considerations of the structure of the
industry should not obscure its massive quantitative expansion, itself
premissed on a similar expansion of, and change in the composition
of, final demand for petroleum: in 1900 oil supplied 3.8 per cent of
world energy demand, by 1940 this figure was 17.9 per cent; and this

represented an increase of oil output by a factor of fifteen.[36]

In sum, the interwar years witnessed the consolidation of an oil industry structure where the moments of capital and the material stages of the oil sector were fully integrated within a few state-supported, transnational enterprises. Within the United States this structure remained relatively stable despite the massive changes in the rest of the economy, as the early consolidation of the oil industry was itself monopolistic. Indeed, the oil industry shifted to a state monopoly capitalist form of accumulation. Outside North America the structure of the industry was premissed on the material base of the interaction of rapidly increasing demand and the locational pattern of new concessions. In turn, competition in the industry – apart from the odd outburst of price competition before 1928 – centred on the struggle for new concessions. This was neither market-based nor intercompany collusion in oligopolistic conditions (though the latter was certainly present); rather it reflected the attempt by the rising hegemonic power to displace the European imperial interests – the United Kingdom and France in the Middle East, the Dutch in the Netherlands East Indies and a free run in Mexico (until the Revolution) and then Venezuela – as the United States pursued a liberal–capitalist, free-trade imperialism.

The interwar era witnessed not only a profound upheaval in the capitalist world economy, but also an interregnum of hemegonic leadership in the interstate system, with US power becoming fully consolidated only through the transformations wrought by the Second World War. The end result was an integration of production, distribution and exchange by the majors of an unparalleled kind. This pattern of accumulation can neither be understood as a technical fix in a large-scale industry with an endemic tendency to oversupply, nor as an allocation of resources via competitive economic processes (albeit distorted by oligopolistic collusion and government interference). Rather, it primarily reflected the emergent role of oil as a strategic commodity, and the resultant impact of state competition (at once economic and geopolitical) in the constitution of the structure of accumulation in the industry.

The Turning Point: The Impact of the Second World War

It is often suggested that the structure of the global oil industry was not significantly altered by the war. Although this judgement is in some ways accurate, it betrays a narrow industry-centred focus which obscures the fundamental transformations underway in the geopolitical economy of oil, changes that were to provide the deepest structuring of the industry through to the 1970s. The war, along with the immediate postwar period, gave momentum to a number of geopolitical and economic changes: the emergence of a basically bipolar global politics, as the Cold War and US hegemony reshaped the world; the erosion of European colonial control, with the ensuing consolidation of the nation-state system in the periphery; and the reconstruction of the world market under US leadership, now increasingly dominated by trans-national forms of capital accumulation. To grasp the role played by oil in the postwar order we must begin by reviewing the character of these wider developments.

The Establishment of Postwar US Hegemony

During the war US government planners began studying policies for the reconstruction of the global order in peacetime. In these strategic deliberations, which included the War and Peace Studies Project of the Council on Foreign Relations, it was recognized that the United States would hold unquestioned power and that an immediate requirement would be a programme of complete rearmament. In the early stages of the war a common assumption was that Germany might control a large part of the European landmass, and therefore Grand Area planning (that is, the design for a liberal capitalist order) encompassed the Western Hemisphere, the British Empire and the Far East. The British Empire was seen as a major threat to this new order, and accordingly lend-lease was carefully rationed, enough to sustain Britain's war effort but not sufficient to enable the retention of her imperial role.[37]

By contrast, whatever its long-term goals, the Soviet Union had one overriding aim – security.[38] And here the content of its aims varied geographically: in Eastern and Southeast Europe and North Korea efforts were made to extend Soviet influence and its system; in Western and Southern Europe, China and the Far East, the obvious areas of US expansion, policy centred on stability, national reconstruction and the co-operation of indigenous communist parties with national bour-geoisies; and in Germany and Austria, where co-operation with the allies was a necessity, the aim was to ensure that after withdrawal the state would not joint the capitalist camp. Manifest conflict centred on Eastern and Southern Europe, where the United States wanted to see the exercise of self-determination and entry into the design of a global market order. Roosevelt's policy attempted to reconcile two divergent paths: a Wilsonian universalism which would involve the breakup of the old power blocs and spheres of influence, along with their replacement by national self-determination within the ambit of a liberal–capitalist international economy. On the other hand, a realpolitik towards the Soviet Union argued for altogether less ambitious goals. In this confluence of conflicting designs the United States had the decisive advantage of absolute economic superiority together with an initial nuclear monopoly (and a continuing leadership).

By early 1946 US policy (now under Truman) had hardened, and an overt policy of containment began.[39] This development reinforced the Soviet consolidation of Stalinist regimes in Eastern and Southeast Europe, but 'Moscow sought until the autumn of 1947 to reach agreement with the United States'.[40] In turn, containment was premissed on a combination of the need for open trading relations to complement the US domestic economy, the desire to contain socialist forces in Europe and the priorities of the domestic military lobby. In 1947 congressional opposition to new loans, along with domestic opinion's reservations over the break with isolationism implied by containment, were both dissolved by the conjunction of the Truman Doctrine and the Marshall Plan – Truman seeing these as 'two halves of the same walnut'.[41] For in 1947 the United States had to reconcile its key objectives in furthering its global interests with respect to Western Europe and the nascent Eastern bloc, averting the impending breakdown in international trade and payments, adhering to the aims of Bretton Woods and the Anglo-American agreements, and resolving the German problem.[42] Coupled with Soviet intentions in the East, these aims generated a series of events which resulted in the division of Europe (the German division being central to stabilizing the European state-system) and the consolidation of Cold War geopolitics.

Whatever its causes, and without minimizing the reality as well as the asymmetry of the conflict involved, the functional aspect of the 'Cold War system' should be noted.[43] Because of both encirclement and internal difficulties in the Soviet order, the Soviet elite found the exploitation of intra-Western conflicts politically useful, in addition to deriving considerable benefits from the division of Europe. In this way, the Cold War system functioned to strengthen internal discipline, and it assisted in the control of Eastern Europe and the international communist movement.[44] Similarly, the Soviet 'threat' soon became 'extremely important for the USA: it successfully mobilized domestic support for America's new international role; created a fixed point of opposition around which to organize the western alliance; and it legitimized American intervention in the Third World'.[45]

On the economic plane, the postwar era of the capitalist world economy saw the emergence of a transnational economic order exhibiting several distinctive features.[46] The first and most striking was the development of a new international division of labour within the cycle of productive capital. Second, industrial development of the semiperiphery began, with regions or states moving from export promotion through import substitution to export substitution. Because of the pace of development, with the associated limits placed on internal demand by polarized class formations, this was highly dependent on the core for export markets, capital goods and finance. Finally, there developed transnational but US-dominated devices for coordination, involving bilateral, multilateral and supranational institutions, free-floating reservoirs (such as Euro-markets) and mini-states (such as Hong Kong). These serviced the new international division of labour. However, this reconstruction of the world market could not have occurred without the emergence of a series of nationally negotiated growth projects in the centres of metropolitan capitalism, among which there were marked regional and international complementarities. And, in turn, this process was facilitated by the policies of reconstruction engineered under the auspices of the Marshall Plan in Europe and the US occupation and reconstruction of Japan.

What generated this growth? The central question of postwar growth has been succinctly expressed by Milward: 'How did a boom whose origins lay in an intensely nationalistic reconstruction of capital goods industries and the national infrastructure turn, without apparent interruption, into an export-led boom in increasingly open economies driven forward by high levels of consumption?'[47]

The transformation of the international political order due to the consolidation of the Cold War blocs and the upsurge of peripheral nationalisms produced a unique political solidarity among the advanced capitalist countries (ACCs) under US hegemony. This made possible the commitments to international trade and monetary liberalization that opened new vistas of economies of scale and specialization. The backlog of technical innovations from the interwar years and wartime, combined with the diffusion of advanced technology from the United States, unleashed large productivity increases. Added to this was a series of national political settlements, or the weakening of trade union organization during wartime and reconstruction, which helped to provide labour quiescence. Demand conditions were highly favourable: nationally these were due to changes in government macroeconomic policies, the growth of built-in stabilizers and the absence of perverse discretionary action; internationally the inherent deflationary bias of the international system – due to the demand-depressing effect of orthodox solutions to payments deficits, which by their success spread deficits elsewhere – was averted by the overall US deficit.[48] The final and often neglected underlying cause of postwar growth was a permissive one. This was the major trend improvement in the terms of trade between the ACCs and the Third World: this was particularly true of raw materials (especially oil): from 1950 to 1963 the 'terms of trade improved by about 25 per cent (from the standpoint of exporters of manufactured goods)',[49] and between 1963 and 1972 remained nearly stable.

The result was a series of complementary virtuous circles of economic growth where 'because of the exceptional opportunities for technical innovation, high rates of investment brought forth high productivity growth rates, whilst strong demand conditions held profits up',[50] while real wage increases did not outstrip productivity growth so that the profit share, confidence and renewed investment were all sustained. One notable feature of these developments was the rapid expansion of investment (and from the mid-1960s trade) between the capitalist metropoles, leading to the formation of a new horizontal (intrabranch) international division of labour with internationally imposed conditions of production. This took place after the nationally based reconstruction and began in the early 1960s. Furthermore, the low level of US export specialization relative to its huge domestic market, combined with national revenue circuits that eluded ties of international financial interdependence, produced what Aglietta terms a 'desynchronization of national conjunctures'[51] wherein national demand management and growth could be pursued relatively independently. In addition, the interconnected growth of the strong economies of Northern Europe (excepting Britain) and the high growth economies of Southern Europe was complemented by the role of the Japanese economy, as the latter (largely impermeable to western exports) contributed to the stability of key developing countries and oil states, themselves import markets for the US and European exports.[52]

This internationalization of capital was intimately bound up with the massive expansion of multinational capital during the long boom. Initially international capital movements were dominated by official (predominantly US) flows and total US capital exports (private and official) peaked in the late 1950s, falling thereafter because of a decline in private outflows from an annual growth rate of 7.8 per cent during 1955–60 to 4.4 per cent during 1965–70. In general, foreign investment expanded more rapidly than national gross domestic products until the mid-sixties, and since then has at least matched gross domestic product growth. From the Marshall Plan to the formation of the European

Community (1958), this internationalization was dominated by US concerns, but from the mid-sixties European and Japanese firms began a process of rapid overseas expansion, while the late-sixties saw the start of a large rise in the foreign activities of western banks. US control of foreign investment fell from 70.4 per cent in 1967 to 58.8 per cent by 1976, and in the late seventies West European and Japanese foreign production increased far more rapidly than that of US transnational corporations.[53]

The integration of these elements can be seen most clearly in the Marshall Plan. Between 1948 and 1952 perhaps one-third of US exports were financed by US aid but recipient nations had to agree to economic policies fostering industrial compromise around productivity improvements, the expansion of trade and the maintenance of financial stability to encourage regional growth. At the outbreak of the Korean War the Marshall Plan 'was transformed into an economic support programme for European rearmament',[54] while the Economic Co-operation Administration (ECA) was renamed the Mutual Security Agency (1951). Indeed, as Clarke has suggested, because of this integration of social, economic and geopolitical aims, the postwar order is perhaps better termed "Marshallism", rather than "Keynesianism" or "Fordism", which strictly describe only elements of the strategy'.[55]

Oil and the Postwar Order: Economic and Military Uses

Oil played a crucial role in these transformations.[56] The postwar boom involved the exploitation of a range of new technologies which had been introduced in the interwar years and given a boost by war production. These included electrical engineering and supply, the motor industry, aircraft production, chemical engineering, oil refining, and the working of aluminium, rubber, plastics, and other materials. In turn, the large-scale displacement of coal by oil, both as an input into production and as a source of energy, was central to this transformation, especially during the 1950s and sixties. This was so in a number of ways. First, oil-based products in the form of fuel for the internal combustion engines of cars, ships and airplanes facilitated the development of transportation. Second, oil products were widely used as lubricants for mechanical parts. Third, oil and gas (or petroleum) became an important input into the petrochemical industry. Fourth, the petrochemical industry contributed to the rise of agricultural productivity, with its production of fertilizers. Fifth, the main growth sectors in the West – cars, consumer durables and petrochemicals –

were all high energy input products. Sixth, oil could heat and power an increasingly industrialized and urbanized society which involved the suburbanization of cities, the switch from public to private transport, the mechanization of housework and better standards of heating. Comparable factors were at work in the Eastern bloc, especially the rapid expansion of heavy, energy-intensive industry. Likewise, industrialization and urbanization in parts of the Third World, combined with highly energy-intensive infrastructure construction, created high levels of energy demand. The upshot of these trends was that during the period 1945–73 the rate of growth of energy use was 5 per cent per year (while the real price of crude oil halved between 1950 and 1973).

The military significance of oil was also on the increase. In this respect, it has been of cardinal importance that the two hegemonic powers of the twentieth century have, for reasons of geography, required extensive naval power. For this too has had profound consequences for the development of the oil industry. By around 1914 it was clear that on technical grounds an oil-burning fleet had many advantages over coal for naval power: 'Not only was it cleaner and lighter [in relation to power generated], but it provided extra speed and range while also permitting refuelling at sea'.[57] Accordingly, despite its greater cost, a secure supply of oil became crucial to the pursuit of military power and competition among naval powers. Moreover, as we saw above, the new oil finds, particularly those in the Middle East, were under the control of areas held by the European imperial powers, primarily Britain. The competition among the the imperial powers, which characterized the years around the First World War, was often bound up with a 'distinct oil diplomacy': oil became a factor in interstate relations, while the trajectory of the industry itself was influenced by state policy.[58] A development of equal significance came later with the complete mechanization of land transport, together with its application to the major land battles of the Second World War, as this further augmented the military imperative to have a secure supply of oil. Indeed, the transformation of logistics during the Second World War – as Mandel notes, 'the first motorized war in history' – led General Marshall to speak of it as 'the automobile war'.[59] To manage this the Secretary of the Interior, Ickes, was appointed Petroleum Co-ordinator for National Defense in 1941, and a Petroleum Industry War Council together with the Petroleum Administration for War were established.

If oil was to be central to postwar economic growth and was recognized as a vital component of military power, Western geopolitical influence in the Middle East had been profoundly altered by the war. This was a potentially serious problem; although US oil provided nearly six-sevenths of Allied oil during the war, it was clear that future supplies to Western Europe (now without Soviet exports) and Japan would have to come primarily from the Middle East.[60] Indeed, US hegemony depended on rapid and stable growth in these areas, and in turn this depended on plentiful supplies of cheap oil. US control over the international oil industry was therefore seen by wartime planners as a key element to their designs for the postwar order, a form of control which was to be 'at the centre of the redistributive system of American hegemony'.[61] Specifically, control over the Middle East oil reserves and the displacement of the erstwhile imperialist power, the United Kingdom, was required in order that US possession of low-cost Middle East oil could underpin European (and Japanese) recovery. The US majors were to play a central role in such a strategy, but the lead in much of the policy-making came from the US state, not the oil industry itself. (Indeed, much of the domestic US oil industry was opposed to the policies involved.)

As noted above, the Second World War strongly reinforced perceptions of oil as a strategic commodity, and this fact was not lost on the US planners. But plans formulated at the end of the war to reduce domestic production of oil – for long-term reasons of security of supply – and increase overseas expansion ran into opposition from the politically powerful domestic oil lobby. None the less, the United States did adopt a determined policy of expansion in the Middle East. As Venn has argued: 'To a large extent two vitally important policies, the conduct of the Cold War against Russia, and the retention of control over the Middle Eastern oil supplies, converged in a general emphasis upon national security objectives, and the retention of western dominance in the Middle East'.[62]

In fact, it was only the weakening of European colonialism during the war and the Cold War unity among the capitalist powers in its aftermath that allowed a volatile situation to be stabilized. An earlier challenge to Western control of the Middle East had occurred in the wake of the First World War, in part inspired by the Bolshevik Revolution of 1917, but British and French imperialism, as well as nationalist–military regimes in Afghanistan, Turkey and Iran contained social and national revolts. On the other hand, the interwar years witnessed a major transition in the power of the major imperialist rivals in the Middle East, Great Britain and the United States. Let us now consider the specific role of oil in the Middle East, as well as the more general place of the region in the development of the world economy and the consolidation of the nation-state system.

The Middle East: From British to US Hegemony

The Middle East Under British Hegemony

During the nineteenth century three major forces contributed to the growth of the economies of present-day Turkey, Syria, Lebanon, Egypt and Iraq: the expansion of the European market for the region's agricultural produce; the emergence of 'strong, centralized regimes in Istanbul and Cairo whose concern with security and more efficient methods of tax collection provided the framework in which population could grow and agricultural output expand';[63] and the disappearance of the plague in the early part of the century, together with later governmental efforts to improve public health. Much of the agricultural surplus generated was appropriated by land-owning classes who either consumed it or devoted it to 'urban political activity'; equally, the rapid growth of trade with Europe resulted in 'infrastructural development and the construction of buildings in the major port cities'.[64] The Gulf had long served as an entrepôt for European trade, and it had produced dates and pearls for export. However, this strengthening of agriculture and comprador interests, combined with the growing commercial, financial and political penetration of the West, meant that in some parts local elites began to consider the need for state intervention in pursuit of national strategies of industrial development.

The British, of course, were already established as the dominant power in the region, but their position was not uncontested. Indeed, the Eastern Question, which centred on the Anglo-Russian competition over the breakup of Turkey, preoccupied statesmen throughout the nineteenth century. While the British were concerned about the route to India as well as protecting their expanding trade with the Levant, the Russians sought control of the entrance to the Black Sea as 'a matter not merely of diplomatic and military importance, but with the growth of Ukrainian grain exports, of economic urgency also'.[65] After Clive's victories gave Britain dominance of the northeast coast of India (the Indian empire was to expand to two-thirds of the subcontinent between 1814 and 1949), control of the Gulf waters became essential in order both to protect the trade route to India – whose market was of increasing import – and to open up the Far East during the Opium War (1839–42).

Accordingly, in 1820 a series of treaties signed with the tribal leaders of the Arab Gulf littoral formed the Trucial system (the *sahel Oman*).[66]

British control as extended throughout the nineteenth and early twentieth century (especially after fears of German and Turkish expansion
into the region were alerted by the proposed Berlin–Baghdad railway),
eventually encompassing the present states of the United Arab Emirates
(1820s), Bahrain (1861), Kuwait (1899), Oman (where French influence
was thwarted by an agreement with the Sultanate in 1891 and with the
Imamate in 1920) and Qatar (1916). Under these arrangements, most
of the Gulf states had 'no recognized legal status within the Empire'.[67]
Premissed on this control, between 1913 and 1922 a series of concessions
were concluded 'by the different rulers who undertook not to award
any oil concession except to a company appointed by the British
government';[68] in addition British banking interests were to be given
special privileges.

The other major development on the peninsula was the rise of the
Wahhabi-Saudi movement and founding of the Kingdom of Saudi
Arabia in 1932. This occurred after Abdul Aziz (crowned Ibn Saud
1926) signed the Treaty of Jidda with the British in 1927 and then used
the provision of mechanized weaponry by the latter to suppress the
Ikhwan.[69] In general, the states of the peninsula only began to define
their borders when the oil companies required to determine the extent
of their concessions. This erection of state organizations, and thus the
incipient extension of the nation-state system to the region, was in part
a particular strategy of territorial control deployed by the British to
facilitate both internal pacification by indigenous elites and administrative coherence for the operation of oil capital. More generally, it
reflected the changes attendant on the consolidation of formal colonialism under the New Imperialism of the imperialist epoch, without taking
a properly colonial form.

On the Persian border of the Gulf, Iran was reduced to the status
of a semi-colony by the incursions of Russia in the north and Britain
in the south; an arrangement formalized by the Anglo-Russian convention of 1907 'which divided Iran up into three respective spheres of
influence; Russian in the north, British – with the oil concession area
– in the south, and neutral in the middle'.[70] The British D'Arcy concession of 1901, which formed the basis of Anglo-Persian's power in
Iran, excluded the five major northern provinces because of Russian
claims in the region. Notwithstanding the Constitutional Revolution
(1905–11), Iran remained weak and was unable to prevent the flouting
of her neutrality by Russia and Britain during the war. Also during the
war the closing of the Dardanelles and the collapse of Russia severely
damaged the Persian economy. The Bolshevik Revolution also deprived
the Shah of his key ally, and a revolutionary movement emerged in

the northern province of Gilan. Meantime, the British Foreign Office sought to establish a semi-protectorate with a treaty in 1919 involving British financial advisers for the Iranian government, retraining of the army and the provision of engineers for railway construction. In this context, by 'April 1920 Soviet power was re-established throughout Azerbaijan'[71] and the Soviets provided support and recognition for the Soviet Republic of Gilan established by the nationalist leader, Kuchik Khan. However, 1921 witnessed the establishment of a military, nationalist regime in Tehran under Reza Khan, and the Gilan Republic, after quarrelling with the Soviets, was to be short-lived. For by October 1921 'Persian forces reoccupied Gilan with Soviet approval, and hanged Kuchik as a rebel'.[72] The military leader, Reza Khan, continued to consolidate his rule and, after abolishing the Qajar dynasty, crowned himself Reza Pahlavi Shah in 1925.

To complicate matters still further, after the defeat of Germany and Turkey, the Allies, and in particular Britain, turned towards a campaign against Bolshevism. This combined with the birth of the Comintern meant that the Soviet Union 'soon found itself committed, in default of other means of defence, to a general diplomatic offensive against Great Britain in Asia'.[73] Contestation occurred in Afghanistan and Turkey as well as Iran. In Turkey the Soviets supported Kemal in 1919 and renewed their commitment after deteriorating Anglo-Soviet relations in 1921, and by spring 1922 the (British-backed) Greek forces had been routed. However, 'not only had Kemal, victory once achieved, no further need of Soviet support, but the chances of a favourable settlement with the west might even be prejudiced by too close an association with the Soviet government'.[74] In Asia, Soviet policy oscillated between support for revolutionary social movements which contested both British imperialism and the indigenous ruling classes and the search for diplomatic alliances with nationalist rulers against Great Britain. Either way the external situation combined with the weakness of the Soviet economy meant that its position was weak. With the retreat of prospects for revolution in the West, together with the advent of the Lausanne conference in the winter of 1922–23, the latter policy was clearly dominant over the former. Thereafter Soviet involvement in the Middle East was all but at an end until the Second World War.

In the imperialist camp, before the First World War the Ottoman Empire had been Britain's principal ally in the Middle East, forming a buffer against Russia. Within the Empire, European economic interests were served through a system of extraterritorial concessions, aptly known as the capitulations. During the war, with the temporary eclipse of Russia due to the Revolution, the British had occupied Palestine,

Syria and Iraq. However, while Anglo-French competition over territory and oil was soon resolved through the Sykes–Picot agreement of
1916 and the San Remo conference of 1920, popular and nationalist
opposition throughout the region – especially in Iran, Iraq, Egypt and
Turkey – threatened Britain's control. Already overcommitted and
subject to severe domestic pressures, imperial hegemony was maintained by Churchill's development of the 'Cairo Strategy' in March
1921, itself a further application of the colonial developments noted
above. As Stivers notes:

> Without peace with Turkey, the costs of Iraqi defense would impose a
> heavy burden on both Britain and the Arab government. The attainment
> of a comprehensive peace was essential also to the development of Mosul
> oil. And this oil development, entailing a flow of royalties to the Iraqi
> government, was in turn a precondition to the overall British strategy of
> building a client state that could pay for itself.[75]

In Iraq the revolts were pacified through the judicious use of (low-
cost) air-power. Simultaneously, a state with a client government was
established under King Faisal in which the British maintained effective
control over military, fiscal and judicial administration. The revitalized
Turkish Petroleum Company operated the Mosul and Basra fields, now
with US participation. Iraq played an important role in the British
strategy for the Middle East: as well as being a key transportation site
and having considerable capacity for crop and cotton production, 'the
possession of northern Iraq was crucial to the control of southern Iraq,
and southern Iraq was a key factor in the defense of the Anglo-Persian
Oil Company's holdings in Persia'.[76] In addition, in time of war the
Iraqi oil fields would be vital to naval power in the region.

As to relations with Turkey, at the Lausanne conference Curzon was
able to separate Turkey and the Soviet Union, secured a regime for
the Turkish straits which suited British interests and 'walked off with
the prize of Mosul'.[77] And Anglo-Turkish relations were consolidated
in 1926 when

> in return for complying with the League's boundary determination [on
> the Iraqi–Turkish border which awarded the oil-rich Mosul province to
> Iraq], the Turks agreed to accept 10 percent of the royalties for a twenty-
> five year period. . . . From this point on, Turkey would once again
> support Britain in the Middle East as a buffer state, throwing her weight
> against a resurgent Russia.[78]

The position of the United States in this period has been the subject of considerable confusion. Two specific problems have confounded a clear appreciation of the role of both US capital and successive administrations: first, it is commonly assumed that the role of the US state in the region was inspired by fear of domestic oil shortages, with the government pursuing a general, national interest; and second, it is routinely assumed that the 'open door' policy entailed an opposition to European power in the underdeveloped world. Neither of these claims survives close scrutiny. As Stivers has shown, the US majors 'were in the vanguard of U.S. penetration into the Middle East'.[79] These companies sought access to control over production in order to complement their already dominant marketing position outside the Western Hemisphere, and they also wanted to compete in the supply of the bunker trade for the merchant marine and the Navy. And with the entry of US interests into the Turkish Petroleum Company the way was open for a general agreement throughout the international oil industry: this was the Red Line Agreement of 1928.

Meanwhile, not only did US administrations repeatedly refuse to side with 'moderate' nationalists – this was true in Iraq throughout the 1920s and in Turkey at the Lausanne conference – against British and French imperialism, they also supported the mandate system and European control in the region. The reason for this was the need both to maintain allied unity against the Soviet Union and specifically to head off any chances of a Turko-Soviet rapprochement. Equally significant was the logic of capitalist expansion during this phase of the world economy. For in a region where states were established on a precarious basis, if at all, and where internal pacification along with the maintenance of secure property rights was routinely challenged, the expansion of US capital required order and stability above all. And while Wilson had announced the Fourteen Points to counter Bolshevik propaganda in Europe and to muster support for the war at home, as in Latin America so in the Middle East, he did not advocate undue haste in transforming the mandates into independent states. Stivers has summarized his position thus: 'until the Arabs had proved to the advanced nations that they respected the sanctity of property and contract and could operate administrative and judicial structures in a manner pleasing to the major powers, they would remain under tutelage'.[80] Overall, US strategy in the region amounted to giving moral (but not material) support to the British, while simultaneously deriving economic benefits from the administrative regimes established by the latter.

US Hegemony in the Middle East

After the war, despite the fact that Britain remained the principal imperialist power in the region (especially in Iran and the Gulf states), the economic and strategic situation was to be radically transformed by a number of new developments: the weakened global position of British imperialism and the growing US presence and role in the region; the rise in power of the Soviet Union; the emergence of Arab nationalism; and the establishment of the state of Israel.

Already in 1933 Ibn Saud faced a financial crisis as income from the *hajj* declined during the Depression and the Japanese began to manufacture artificial pearls. At the same time, the Kingdom received a large payment when Socal successfully bid for an oil concession, and oil production commenced in 1938.[81] In 1943 Saudi Arabia was declared eligible for lend-lease on a State Department initiative, and this move was supported by those US majors, Socal and Texaco, which had Saudi concessions and had lobbied for such a policy for two years. A proposal to set up a state-controlled Petroleum Reserves Corporation (PRC) that would have owned large amounts of Saudi oil was defeated by opposition from Exxon and Mobil. A further proposal for the PRC to build a government-owned refinery in Saudi Arabia was aborted due to company and isolationist congressional opposition, with the outcome that Socal and Texaco decided to build a private one.[82] The resistance of Exxon and Mobil reflected their hostility to state intervention which would have given advantage to their competitors, Socal and Texaco. So, a less discriminatory strategy was needed that would none the less guarantee heightened US control over Middle East oil. In 1944 an Anglo–American International Petroleum Commission was proposed which was to have recommended 'production and exploration rates for the various concessions in the Middle East . . . [to prevent] the disorganization of world markets which might result from uncontrolled competitive expansion'.[83] This suggestion fell too, essentially because of opposition from US domestic producers who feared that it might lead to the exploitation of foreign reserves at the expense of domestic output. Eventually, US control over Saudi oil was secured through the abrogation of the pre-war Red Line Agreement which had been preventing Exxon and Mobil's expansion in the region. In 1946 Socal and Texaco (Aramco), with their large Saudi concessions, were crude-long, capital- and outlets-short and unconstrained by the Red Line Agreement, and accordingly they successfully invited Exxon and Mobil to join the Aramco consortium.

In Iran, meanwhile, British and Soviet troops had been stationed to guard the oilfields against German advances, with each agreeing to remove their forces shortly after the end of hostilities.[84] However, with the British evacuation from southern Iran underway, in the winter of 1945–6 two autonomous republics were declared in the north, one in Azerbaijan, the other a Kurdish People's Republic. And when the Iranian government attempted to suppress these movements Soviet troops took preventative action. Under pressure from the Tudeh Party, the Iranian regime began negotiations with Moscow in which the latter demanded that the two northern provinces be accorded an autonomous status and that a joint Soviet–Iranian oil company be formed. The United States encouraged the Iranians to place the matter before the United Nations Security Council while themselves adopting a tough line in bilateral talks with the Soviets. (US companies were also aiming for an oil concession in northern Iran.) In April 1946 a Soviet–Iranian agreement was concluded and Soviet forces withdrew, but at the end of the year the Tudeh Party was expelled from the ruling coalition and the Majlis refused to ratify the oil concession. (Along with Stalin's proposal in August 1946 for joint Turkish–Soviet defence of the Dardanelles, events in Iran contributed to the hardening of Washington's nascent policy of containment, as well as the permanent stationing of the US Navy in the eastern Mediterranean.)

Another postwar problem that the United States faced concerned the attempt to displace British influence in the Middle East, a task rendered complex by the fact that all of the non-Red Line Agreement concessions involved BP. In particular the sterling–dollar oil issue of 1949–50 proved a source of conflict. The origins of this dispute lay in the attempt by dollar-short Britain to prevent sterling area countries purchasing dollar-denominated oil which, because of the conditions of oversupply, resulted in a strong protest by the US majors. (The Gulf states held their sterling balances in London.) This issue was finally resolved through complex interstate negotiations and a series of swapping arrangements – and in 1950 the Korean War eradicated the 'oil surplus' that lay at the heart of the problem. However, Britain's imports continued in effect to be sourced by sterling-denominated oil, and this in turn played a significant role in the Suez Crisis of 1956. In sum, despite the abortive attempt at an Anglo-American oil agreement and general US–British competition (itself moderated by the joint concern to exclude the Soviet Union from Iran), the US position in Saudi Arabia and the British in Iran – and after the CIA-aided coup against Mossadeq the US share of the Iranian concessions rose from nil to 40 per cent – meant that Western interests in the Middle East were secure.[85] Indeed, the joint Anglo-American position was such that the British-organized boycott of Iran's oil could be offset by increasing crude production in Iraq, Kuwait and Saudi Arabia.

In addition, a new complicating factor (of great long-term significance) was introduced by the establishment of the state of Israel in 1948. Already in the interwar period the deteriorating situation in Europe resulted in an increase of Jewish migration to Palestine, encouraged by the Balfour Declaration of 1917. After the Second World War this migration increased still further. The Truman administration was deeply divided over what attitude to take in the UN debate (1947–8) on the partition of Palestine, with Truman's stance being heavily influenced by domestic electoral considerations. The problem, as officials were well aware, was that support for Israel could jeopardize relations with the Arab states. The administration's thinking on this issue has been well expressed by Kupchan:

> The State Department feared that the establishment of Israel would jeopardize American relations with the Arab world and potentially lead to the interruption of the flow of oil to Europe – an energy source essential to the viability of the Marshall Plan. Many officials also argued that the regional instability caused by partition would serve as an opening for the Soviets. The Defense Department voiced concern about losing military access to the airfield in Dharhan, Saudi Arabia.[86]

None the less, with the British abdication, the United States did support the state of Israel, soon becoming its sole protector. To start with, during the 1950s, 'the U.S.–Israel relationship was decidedly uneasy'.[87] But after the upsurge of radical nationalism in the 1950s, and especially after the 1967 Arab–Israeli war, successive US administrations 'increasingly came to accept the Israeli thesis that a powerful Israel is a "strategic asset" . . . serving as a barrier against indigenous radical nationalist threats, which might gain support from the USSR'.[88]

By the mid-fifties, with the Shah restored in Iran, Eisenhower sought to strengthen the US position in the Middle East through the creation of anti-Soviet alliance. And so, through the orchestration of Dulles, a series of bilateral pacts with Turkey and Pakistan were cemented by British, Iraqi and Iranian membership into the Baghdad Pact of 1955. If the United States saw the pact as a means of collective defence (as well as providing a cover for considerable internal repression, especially in Iran), the British were also concerned to defend their interests in Iran as well as strengthen Iraq vis-à-vis Nasser's Egypt. Eden saw Nasser as a threat to Britain's great power status which rested upon leadership of the Commonwealth, the sterling area and access (via the

Suez Canal) to sterling-denominated oil.[89] In 1955 Nasser had signed an arms deal with Czechoslovakia (followed by Soviet arms deals with Syria and Yemen in 1956), while Khrushchev also promised Soviet finance for the Aswan Dam and agreed to purchase Egyptian cotton. After the Suez débâcle, the Nasser regime became more radical, and in early 1957 a radical regime emerged in Syria. In response the United States Congress passed a resolution – the Eisenhower Declaration or Doctrine – authorizing the president to take action against any 'armed aggression' by 'International Communism' which threatened states in the region.

Lacking oil of their own, Egypt and Syria were none the less of crucial importance as most of Europe's imports from the region crossed theiry territory, with some 65 per cent of Middle East oil transported in this manner in 1958. The formation of the United Arab Republic in 1958, together with General Kassem's coup against the Hashemite monarchy of Iraq in 1958, resulted in Britain and the United States sending troops to Jordan and Lebanon, an action which represented a significant escalation of superpower struggle in the region.[90] After the abrupt ending of the Baghdad Pact consequent upon the fall of the Hashemites, August 1959 witnessed the US formation of the Central Treaty Organization with Turkey, Iran and Pakistan as fellow members.

Finally, the Suez crisis of 1956, in which a combined Anglo-French-Israeli invasion of Egypt took place on the pretext of forestalling an Israeli–Egyptian conflict, signalled the demise of British power in the region and the final ascendancy of the United States, itself confirmed by the dispatch of marines to Lebanon. Under the Kennedy and Johnson administrations the United States increased its military assistance and arms sales to Israel and Saudi Arabia, while the Soviet Union continued to supply Egypt, Iraq and Syria. By the late 1960s, superpower involvement had been augmented by the Arab–Israeli War of 1967, the consolidation of oil production and reserves in the Middle East (Arab states had attempted to limit the West's supply of oil in response to US support for Israel), the British defence review of 1968 which announced her intention of withdrawing 'east of Suez' by 1970, and the emerging US–USSR rivalry in the Indian Ocean (in 1968 the United States deployed Polaris missiles which could reach the Soviet Union).[91]

The Twin Pillar Policy: Iran and Saudi Arabia

If the demise of French power in the Middle East was signalled by its exclusion from the Baghdad Pact, the Suez crisis further strained relations between the United States and Western Europe. The Arab–Israeli Six Day War (1967) was perhaps of even greater moment. As a result of this conflict the United States displaced France as Israel's largest arms supplier, while the Arab world was further divided by its defeat. The Soviet Union, on the other hand, gained a base in Egypt in return for helping to rebuild the latter's military forces, broke off diplomatic relations with Israel and offered support to the Palestine Liberation Organization. Thus the ascendancy of US over European interests within the West coincided with the injection of superpower competition into the Arab–Israeli antagonism. However, as a result of the accelerating tide of decolonization throughout the Third World, the United States could not simply replace European power. Therefore, in marked contrast to its position in Europe and Southeast Asia, United States power in the region depended on close alliances with Iran and Saudi Arabia rather than a direct presence of forces or *matériel*. Indeed, Halliday has pointed out that something like a division of labour pertained between Iran and Saudi Arabia in the US strategy. Within the Arab world, there emerged

> a more active Saudi role in Arab politics which the defeat of Egypt in 1967 and the British withdrawal from the peninsula in the late 1960s encouraged. Saudi Arabia, with growing US backing, saw itself as the leader of the Arab world and the guarantor of 'stability' in the Middle East as a whole and in the Gulf in particular.[92]

Alongside Israel, however, Iran was the key guardian of Western interests in the Gulf: it bordered the Soviet Union and it was by far the largest and most advanced state in economic and military terms. But Iran could not serve Western interests alone, in part because of the greater dependence of the latter on Saudi oil and also because any Iranian intervention in regional Arab disputes would almost certainly provoke a nationalist reaction. Thus, as Halliday avers: 'Iran had clearly appropriated to itself the tasks of patrolling the strategic waters of the Gulf and the approaches in the Indian Ocean . . . [and] Iranian money and covert Iranian military missions operated in certain Arabian peninsula states – Oman and North Yemen. But the dominant power, politically and militarily, on the Arabian peninsula itself remained Saudi Arabia'.[93]

The application of the Nixon Doctrine to the Middle East, the twin pillar policy of 1969, thus coincided with British withdrawal (Kuwait gained independence in 1961, Bahrain, Qatar and the United Arab Emirates in 1971) and heightened Soviet involvement in search of influence and bases, a turn accentuated by the desire to 'pre-position large quantities of heavy weapons in North Africa'.[94] In turn, Nixon's policy of building up regional clients suited oil-rich Middle East states such as Iran and Saudi Arabia admirably, as they could recycle dollars earned through oil production for US-supplied weaponry. It was this, together with the fact that while Moscow could supply arms it could neither secure peace nor negotiate the return of the lost territories, which gave the United States such a decisive advantage in the region (a point confirmed by the October War in 1973). Indeed, in 1972 Sadat demanded the removal of Soviet military advisers and troops from Egypt, and the Soviet Union countered by increasing their support for Iraq and Syria: military support to Syria was significantly augmented, while in Iraq 'the Soviet Union undertook to help to distribute oil from the Iraq Petroleum Company field at Kirkuk, which the Iraqi government had at last nationalized on 1 June, and to bring the North Rumaila oil field into large-scale production'.[95] Despite such commitments, as well as Soviet support for the Palestine Liberation Organization, Kupchan has rightly noted that 'Arab and Western economic interests converged after 1973 in a way that could only reinforce the position of the United States in the Arab world'.[96]

Postwar Oil and US Power

Alongside its general advances in the Middle East, the United States adopted a range of policies both to consolidate its control over the postwar oil industry and to secure the role of the latter within the project of global leadership or hegemony. Specifically, US policy relied upon the allocation of Marshall Plan funds to pay for dollar-denominated Middle East oil imports, while restricting their use for financing any increase of European-owned refinery capacity (a similar policy was adopted by the occupying authorities in Japan). In the case of Western Europe, the majors adopted a determined policy of undercutting the price of coal, thus increasing reliance on dollar-oil and thereby strengthening US hegemony. To this end, the procurement policies of the ECA resulted in the Middle East displacing the Western Hemisphere as the primary source of Western Europe's oil: in 1947 Western Europe received 43 per cent of its crude from the Middle East, 66 per cent in 1948 and by 1950 some 85 per cent; and between 1947 and 1950 Middle East, dollar-denominated oil production increased by 150 per cent while non-dollar oil rose by 85 per cent.[97] In a complementary manoeuvre, the ECA pressured the majors to reduce the price of crude supplied to Western Europe and in particular to end the old pricing system. Two base points were established, the Gulf of Texaco (the old one) and Ras Tanura in the Arabian Gulf, with a price regime in which for 1946 the Gulf price was $1.56 per barrel whereas the Ras Tanura price was only $0.9 per barrel (still triple total production costs).[98]

This restructuring of the industry fed the massive expansion of demand for oil between 1945 and 1973. World output expanded nearly eightfold from 1940 to 1970 with the Organization for Economic Co-operation and Development (OECD) taking 68 per cent, centrally planned economies 17 per cent and the Third World 15 per cent of total demand in 1972. Total oil demand was divided by end-use as follows: 39 per cent in the transport sector; 7 per cent for non-energy uses; and 54 per cent for fuel consumption. As a percentage of total world energy demand, oil (and natural gas) provided 17.9 per cent (4.6 per cent) in 1940 and 46 per cent (18.6 per cent) in 1978. Indeed, by 1973 total oil imports into the OECD were seventeen times their level in 1950. And finally, the displacement of coal in Europe was registered by the dramatic fall in the coal to oil ratio.[99]

At this stage in the industry, control moved beyond an administered market and price control to a sophisticated mechanism involving prede-termined growth rates which were kept to a remarkable degree of accuracy, a series of production-control strategies that were often oper-ated with the aid of the US State Department and a complex price-matching system.[100] Combined with US prorationing, this enabled a doubling of prices between 1946 and 1948 and the profit rates of the majors show a high plateau from the late 1940s to the late 1950s: The Iraqi Petroleum Company's profit rate for 1952–63 was 56.1 per cent per year, the Iranian consortium from 1955 to 1964 reaped 69.3 per cent and Aramco over the period 1952–61 garnered 57.6 per cent.[101] In addition, the US majors were taking advantage of the tax concessions provided by the Treasury – in effect a disguised (from the Israeli lobby in the United States and others) form a grant from the US government to the Saudi regime – and in the early fifties the tax 'loss' from Aramco's operations in Saudi Arabia was some $50 million per year.[102] Thus despite the formal independence of the relevant states, the regional hegemony of the United States together with the virtually complete dominance of oil capital over the oil concessions meant that Western control over Middle East oil – itself unified under the banner of Cold War – was complete.

However, this international order contained within itself and more importantly in its relations to wider developments in world politics and the long boom the seeds of its own destruction. According to Gisselquist, the share of US reserves in the non-communist world's total fell over the period 1945–57 from 39.8 to 13.9 per cent while those of the Middle East increased from 37 to 71.3 per cent.[103] Over a similar period, 1948–59, the control of the majors over US imports fell from 74 to 37 per cent as a result of the activities of the independents; and as a percentage of domestic production imports rose from 6 to 14 per cent.[104] Simultaneously, there developed a temporary global surplus of oil production. The majors and the US government welcomed this to the extent that it gave them increased leverage over the producer-states: however, the growth of imports controlled by independent companies also threatened the structure of the domestic industry and increasing imports undercut US prices and threatened prorationing. Prior to the rise of the independents the price structure of the domestic industry, in which profits were largely generated upstream in the production of crude (partly to keep refinery margins tight and thus block entry by newcomers), together with the majors' control over imports and the prorationing of crude, enabled the latter to control the US sector. As a result of pressure from small and medium domestic producers who faced rising production costs, fears of rising US dependence on Middle East oil (rather than the 'secure' supplies of Latin America) and a concern over the trade balance and its impact on the dollar, Eisenhower imposed statutory import quotas in 1959 after two years of ineffective voluntary controls.[105]

These quotas, which were increasingly ineffective as well as counter-productive from the early 1970s and were abandoned in 1973, had three principal effects. First, they strengthened the multinationalization of the independents, because before the quotas US independents, encouraged by fiscal incentives and the depletion allowance, had acquired cheap foreign reserves in order to supply the domestic market. And by the late 1950s the independents owned some 30 per cent of the overseas reserves of US companies.[106] With the imposition of quotas these independents began to market their newly acquired crude in Western Europe and Japan, thus putting downward pressure on prices and further displacing coal. This independent activity helped to feed the massive increases in oil consumption during the 1960s: in Western Europe consumption rose by 300 per cent and in Japan by 500 per cent. Second, the quotas held US crude and thus refined products above world prices, thereby damaging the international competitiveness of domestic, energy-intensive, manufacturing industry,[107] Third, the quotas accelerated the decline of US reserves and so strengthened the position of the Middle East producer-states as the United States became increasingly dependent on Middle Eastern imports.

In addition to these developments, the Soviet Union exported crude oil to Western Europe from the mid-fifties to the mid-sixties, while also developing joint ventures in the Third World. Although the scale of these interventions was small, they were symbolically and materially important. The Soviet Union offered significantly better terms than the majors in India and Ghana, for example, and imported petroleum from Iran, Afghanistan and Iraq. And while Soviet exports to Western Europe, sold below world market prices, were almost certainly motivated by a desire for foreign exchange to finance imports, the United States persuaded NATO 'to oppose any long-term or large-scale marketing agreements between the USSR and Western Europe because of the security implications of any potential dependence by the latter'.[108] From the perspective of the Europeans, and especially the Germans and the Italians, the United States was using fears over 'security' in order both to prevent them gaining independent access to petroleum outside of the control of US firms and to wage economic warfare against the Eastern bloc. That this was so was illustrated by the US-enforced NATO Council decision to restrict West German sales of large diameter pipe to the Soviet Union for its Druzhba oil pipeline to Eastern Europe.[109]

The arrival of the independents transformed the power relationship between the majors and the producer-states in other ways too. The independents began to break the majors' monopoly of control over the circuit of capital in the oil sector, and they transformed the bargaining position of the producer-states in so far as independents were often dependent on a single state for their supplies in contrast to the interterritorial control of the majors. For the producer states began to establish national oil companies (NOCs) – in Iran (NIOC, 1951), Kuwait (KNPC, 1960), Saudi Arabia (Petromin, 1962) and Iraq (INOC, 1964) – which preferred deals with the European NOCs (such as ENI of Italy) and the independents (such as Occidental). Notwithstanding the steady displacement of European coal by US-controlled Middle East oil, the governments of Western Europe (and Japan) encouraged the construction of domestically owned refinery capacity: in part this was designed to ease the balance of payments implications of dollar-denominated oil. Furthermore, Japan and France imposed controls on the majors; Italy (via ENI) and France (via CFP) sought to challenge US control over the industry; and Sweden, followed by Germany, turned to imports of Soviet oil. These developments extended price competition and so deepened the downward movement on crude prices (by 1970 the real price of crude was only 40 per cent of its 1950 level).[110]

The Emergence of OPEC

The formation of OPEC was related both to the postwar phase of decolonization (with its associated rise of Third World nationalism) and to the attempt of producer-states to resist a reduction in the 'posted' or tax-reference price of crude which resulted from the competitive developments discussed above. These changes all contributed to the emergence of a limited market in oil outside the control of the majors. Meanwhile, as the United States moved to a position of being a net importer, the decline of Canadian exports combined with the exploitation of new producing areas produced profound shifts in the oil market. Between 1960 and 1969 the US balance of trade in oil (and oil products) was -0.9 billion, by 1970 it was -1.5 billion, -6.5 billion in 1973, -22.0 billion by 1974 and in 1978 -42 billion; these sums corresponded to net oil imports expressed as a percentage of domestic production of 46.5 per cent in 1973, 74.2 per cent in 1979, falling back to 38.4 per cent by 1983.[111] Moreover, during the 1960s the Middle East displaced North America as the largest oil-producing area in the world.

Until the end of the 1960s the posted price of oil was set by the majors and this, in turn, determined the tax revenues of the producer-states. Indeed, in the mid-sixties a mere 8 per cent of the final cost of petrol to the Western consumer was accounted for by the taxes or royalties going to the host governments; the bulk, some 55 per cent, went to the governments of the consumer-nations in the form of indirect taxes.[112] Also during the sixties there was a general reduction in the real price of energy, with the result that the rate of surplus profit generated in the oil sector fell. This was exacerbated by the movement of production into new areas that required large-scale capital investments. Al-Chalabi explains the consequences of these developments:

> at a time when the per barrel government take in the Middle East was constant, that 'take' in new producing regions (e.g. North Africa) was fluctuating with varying 'realised market' prices of crude sales on the free market . . . It was, therefore, in the interests of the independent oil companies to sell oil at low prices in order to reduce the tax payable to the host governments . . . [and these] price cuts made by the newcomers threatened the dominant market position of the major oil companies . . . [and so they] undertook price cuts in 1959, followed by unilateral price cuts of Middle East oil prices in 1960.[113]

Thus was OPEC born in resistance to the unilateral reductions of posted prices: thereafter, concerted opposition led to a convergence of the systems of remunerating host governments across the world, such that 'the period between 1965 and 1969 witnessed a relative stability in the market place, albeit at a low level'.[114]

Conclusions

As we have seen above, the transnational consolidation of the oil industry was an interwar phenomenon. (This was also true of other minerals which saw the consolidation of Western control during the era of colonialism.) Indeed, the emergence of this structure not only involved the separation of monopoly from non-monopoly capital but also the transformation of the majors into semistate monopolies – that is, they became the privileged recipients of concessions from host governments and of tax subsidies from their parent governments. For foreign operations, tax concessions took the form of allowing the increased payments demanded by the producer-states to take the form of taxes rather than royalties, which could therefore be offset in full against domestic liabilities. The State Department also rendered assistance through its attempts to control entry into the industry through diplomatic activity. This overall dominance reflected the relatively weak position of the oil states in the US-dominated international system, the economic power of Western capitalist interests in the oil industry and the role of oil as a strategic commodity.

The Second World War, however, transformed the interwar setup in a number of ways. The basic material change is simply stated. In 1943 'oil from the Caribbean and the Gulf of Mexico accounted for approximately 73 per cent of the European market whereas oil from Iran and Iraq made up around 23 per cent',[115] the Soviet Union supplied more of Western Europe's oil than the Middle East at the outbreak of the war, and Japan was largely dependent on imports from the United States before Pearl Harbor. By the early 1970s, over 70 per cent of EC imports came from the Middle East (48 per cent) and North Africa (23 per cent), 90 per cent of Japan's oil was sourced by US firms from the Middle East and the United States was increasingly dependent on oil imports itself. This massive alteration in the economic and strategic significance of the Middle East coincided with the shift of regional hegemony from Britain to the United States in the context of a global Cold War rivalry, the reunification of the world market under the dominance of US capital and the consolidation of the nation-state system. That relations among the majors remained oligopolistic and collusive – albeit with a greater US presence – obscures the fact that a quite different form of control now operated. As Painter has ably shown, 'by 1941, the first "Anglo-American petroleum order" had broken down under the impact of depression, world war, and the growing ability and desire of producing countries to control their economic destiny'.[116]

After a number of false starts, by 1948 a new order had emerged. US policy in the Middle East underwrote:

> a series of private arrangements among the major oil companies which promised to accelerate the development of Middle East oil and thus reduce the drain on Western Hemisphere reserves. At the same time, the Truman Doctrine, with its call for the global containment of Communism, provided an ideological basis for maintaining the security and stability of the Middle East.[117]

And as we have seen, this was at once an economic and a strategic undertaking in which the demands of US hegemony played a central role. The relations sustained between the US state and the international oil industry transformed US oil capital from a powerful interest in US foreign policy-making into a key arm of US world power.

None the less, the 1950s and sixties witnessed a series of general changes within the international oil industry – the rise of the independents, the differentiation of the Western European oil market, the start of limited Soviet participation in the world market, the relative decline of US production and the emergence of OPEC – which weakened both the control of the majors over world oil and the position of the United States with respect to the producer-states of the Middle East. Accordingly, the level of collusion and the degree of concentration in the industry decreased, leading to an associated decline in the monopoly power of oil capital. At the same time, US import-dependence increased as its economic position faltered. Further, new forces (the independents, the Soviets and the governments of Western Europe and Japan) shifted the relative market shares, while the emergence of OPEC and the NOCs challenged the majors' ability to unilaterally determine the cost structure of the industry. (Before long the ownership structure would also be challenged.) This meant that the complementary interterritorial control over supply and the price-fixing management of demand was partly eroded.[118] In chapter 4 I explore the consequences of these changes in the wider context of postwar development.

4

The Economic Crisis, Oil and US–OPEC Relations: The 1970s

Introduction

In 1973/4 the 'company-centred regime was destroyed through the exercise of producer's state power in a tightening market'.

Keohane, *After Hegemony*[1]

The postwar oil regime characterized by private corporate control (although hardly a free market) was shattered in the early 1970s.

Krasner, *Structural Conflict*[2]

The events of 1973/4 were 'the culmination of a broader producer challenge to the established rules of the international oil game'.

Bull-Berg, *American International Oil Policy*[3]

The functional relations between political, military and economic power 'have been disturbed – one might even say destroyed – by the recent use of oil as a political weapon . . . [which] could reduce Japan to the status of satellite, a dependency of the oil-producing nations'.

Morgenthau, *Politics among Nations*[4]

The action of OPEC 'was undoubtedly the greatest forced redistribution of wealth in the history of the world'.

Gilpin, *War and Change in World Politics*[5]

The first oil crisis was an event without parallel in the annals of international commerce. Never before had the world witnessed an exercise of monopoly power on such a scale and with such success.

Fieleke, *The International Economy Under Stress*[6]

The oil shocks appear to have ended the ear of high growth and full employment.

Yergin, Global Insecurity[7]

The King of Saudi Arabia, 'Faisal did probably more damage to the West than any other single man since Adolf Hitler'.

Washington Post, Editorial[8]

These quotations, with their quite typical degree of hyperbole, illustrate the conventional wisdom in the West regarding OPEC and the events of the 1970s. For much orthodox argument has stressed the way in which the declining weight of US oil production within the world industry, together with OPEC's assertion of control over price setting and nationalizations, undermined the West's oil order, and with it a vital prop to US leadership. Equally taken for granted was the related claim that the (exogenous) oil price rises were the cause of the economic recessions of the 1970s and early eighties. Underpinning judgements of this kind is a familiar combination of neoclassical models of the economy which stress the more or less stable, long-run capacities of market economies for growth alongside an essentially realist understanding of the distribution of power in the international system. But in fact the economic crisis of the 1970s was primarily the result of development *within* the world economy, in conjunction with the uncertainties introduced by the associated challenges to US hegemony. In turn, the response of the United States to these changes provided a large part of the context for the oil price rises. For the disjunctures within the oil industry itself were a result of far-reaching changes in both the circuits of oil capital and the strategies of the US state in the management of its global hegemony, as well as the actions of the OPEC states themselves. Understood thus, these price increases (in which OPEC itself was merely one player among several) were the symptoms of a wider crisis, at most playing a catalytic role in the overall process. Placing the role played by OPEC in this broader context of US hegemony and the crisis of the long boom reveals an altogether different assessment than those cited above.

In this chapter I shall present an alternative specification of the general conjuncture of the 1970s as well as the particular role played by changes in the world oil industry. For as a consequence of the dual determination of accumulation in the oil sector – that is, the conjoint impact of capitalist competition and state policies – its trajectory has been routinely structured by developments in both the world economy and the composite strategies of the relevant states. In turn, transformations in the control and price of oil have impacted on the broader economic trends and changes in the position of US hegemony. In this way, US attempts to manage its global position through policies directed towards the oil industry have been both consequence and cause of these wider determinations. Accordingly, I shall consider (1) the character of the general economic crisis of the 1970s, paying special attention to the response of the United States; (2) the origins and impact of the momentous changes in the oil industry which seemed to dominate the years 1973–4, situating the role of OPEC within the overall pattern of events; and (3) the attempt by the United States to restabilize both the oil industry and its economic and geopolitical position through bilateral relations with Saudi Arabia and Iran on the one hand and by sponsoring the multilateral International Energy Agency (IEA) on the other.

The Economic Crisis and the World Economy

The character of the economic slowdown in the late 1960s and the subsequent recession is germane to an understanding of the interrelations between oil and hegemony because the recessions of the mid-seventies and early eighties are very often ascribed to the oil price rises that immediately preceded them, and also because those price rises are taken to be exogenous. This general consensus is of considerable ideological and political significance. Said's account in *Covering Islam* has shown how viewing the Middle East through the lenses of the West's economic and strategic interests produces a public discourse which reifies and demonizes the reality of which it purports to speak.[9] In turn, this ideology plays a legitimating role in the design of strategies to reassert the West's hegemony in the region. Within the West itself, the depiction of economic problems as an external force serves a clear ideological purpose for the advanced capitalist countries in so far as it hinders any genuine understanding of the deep crises these societies and their global order face. For in reality, the crisis was neither exogenous nor primarily concerned with the oil price rises.

The extended economic crisis of the 1970s and early eighties was the complex result of fundamental transformations taking place in all three moments of the circuit of capital. These were congealed into a crisis by the particular, conjunctural features of the early 1970s: it was a partial awareness of this conjuncture that constituted the rational kernel of the McCracken Committee Report.[10] Thus, the synchronization of the trade cycle in the major OECD countries during the period 1969–73

together with the general loose monetary policies pursued in the United
States during the Vietnam War and generally after the breakup of
Bretton Woods in 1971 'contributed to an extremely rapid and
synchronised boom, which in conjunction with a number of important
shocks led to an outburst of world-wide inflation almost unprecedented
in peace-time, driven primarily not by wage increases but by price
increases, in which primary products and speculative elements played
a leading part'.[11]

However, the malaise was not merely the result of policy mistakes
nor even of the congealing of the factors detailed above.[12] Inflationary
pressures were generated prior to the events singled out in the
McCracken Report as a result of a shift in the distribution of income
to profits, a slowdown in the growth of real wages, a disturbance
in the wage structure and a reorganization of production and union
organization in the late 1960s.[13] More important still, it was not price
expectations and inflationary psychology that were the causes of
inflation, but rather real wage resistance in conditions of an expanded
tax burden, worsening terms of trade, a lax monetary policy and
above all reduced productivity growth.[14] In the seven major advanced
capitalist countries (ACCs), the growth of labour productivity
decreased from 3.7 per cent per year in the period 1967–73 to 1.4 per
cent per year in 1973–80.[15] Finally, at least as important as the initial
acceleration of inflation was the question of why it proved so endemic
during the recession and varied so much across different countries.

Postwar Growth

Now, based on the US-sponsored reconstruction of the world market,
postwar capitalist development witnessed a growth of direct foreign
investment of productive capital, primarily between capitalist metropoles but later involving the newly industrializing countries (NICs),
which restructured international economic relations. These processes,
along with the more general patterns of economic expansion, were
highly uneven in their effects. The most basic transformation wrought
by these developments was the increasing mobility of capital across
time and space.[16] In addition, the core economic processes of the
ACCS became more integrated through the operations of transnational
corporations and transnational banks. Finally, the frontiers of
advanced, industrial capitalist production expanded to the NICs and
beyond. Initially based on national and regionally ordered growth
patterns, together with the postwar stabilization of class struggle,

profitable accumulation was consistent with a balance between productivity growth and regularized increases in consumption norms. With
the recovery of Western Europe and Japan, however, capital accumulation outstripped these markets and foreign direct investment began
to accelerate. Carried through under the auspices of the transnational
corporations, this investment was aimed at the penetration of foreign
markets in conditions of oligopolistic competition: US direct investment
was initially a defensive move to beat European and Japanese trade
barriers, but as accumulation in these regions expanded, a series of
trade rounds negotiated tariff reductions. The regular overaccumulation
of capital and the associated uneven development implied that this
growth was necessarily unbalanced within industries, between departments and among states.

For a while this could be accommodated through the equilibrating
mechanisms of the international financial system based on the liquidity
provided by the hegemony of the dollar. As Itoh explains:

> As long as U.S. finance capital maintained its competitive export superi-
> ority, the spending of international dollars brought favourable reper-
> cussions on effective demand for U.S. exports. . . . [But despite this
> and military spending] U.S. equipment investments . . . tended to stag-
> nate since the end of the 1950s. . . . In contrast, West European (in
> particular West German) and Japanese capitalism showed a rapid recon-
> struction and growth, continuing active equipment investment until the
> 1960s. . . . Besides, the burden of military expenditure was small in
> West Germany and Japan. Relatively cheap labor power was available
> from agricultural areas in Japan and from nearby countries of the Com-
> mon Market for West Germany.[17]

By 1971, however, the US trade balance went into deficit and simultaneously the US captial account was hit by overseas government
expenditure and continued private foreign investment. This, together
with the slackening pace of accumulation within the US market, made
the Bretton Woods system increasingly difficult to sustain. Nixon's
break with gold and devaluation in 1971 removed these constraints on
the United States.

Meanwhile, the relative economic decline of the United States produced instability in the fields of monetary, trade and energy policy, while
the growth of private, international money capital – beyond national
control – was originally the consequence of the persistent US payments
deficit. Thereafter, the oil surpluses augmented this process, as did the
endogenous credit creating mechanisms of the debt economy itself.
Thus the major sources of the surpluses in the international banking

sector were 'the dollar deficit incurred by the US on the one hand, the oil surpluses of the OPEC countries on the other. The effect of the dollar deficit predominated from the late 1960s to 1973, and again from 1977 to 1979; the oil surpluses from 1974 to 1976 and from the latter part of 1979 to [1982]'.[18] The consequent internationalization of financial capital facilitated and was itself encouraged by these processes. In the long run, and following the breakup of Bretton Woods, three components became particularly important: the internationalization of domestic currencies through the development of markets premissed on floating exchange rates; the internationalization of the operations of bank capital; and the growing internationalization of capital markets.[19] In turn, this disjuncture of global financial circuits from national control made adjustment to the fall in profitability and productivity slowdown extremely difficult. The result was that profit trends in manufacturing declined before 1973, a consequence of a fall in the profit share and the rise in the capital to output ratio.[20]

In terms of the balance between production and consumption, during the long boom realization dilemmas of capital accumulation were initially resolved on a national and regional basis, as can be seen from the fact that the share of exports in gross national product reached its historical low for the ACCs in the mid-sixties.[21] From this low point, international trade increased as capital sought productivity gains through trade or through the utilization of cheaper wage zones in the semiperiphery. With the onset of the recession, the competitive struggle between the major blocs – primarily the United States and Japan and to a lesser extent the European Community – for the lead in the new technologies of microelectronics, biotechnology and energy resources intensified. In the old 'Fordist' sectors (steel, autos, electronics) Western European and Japanese producers undercut the United States; and production processes began to be divided, with large links of the commodity chain being moved to semiperipheral countries, including the Eastern bloc.[22] Throughout, the international restructuring of capital continued apace along several interconnected dimensions. Within the ACCs, diverse strands of capital restructuring emerged (take neo-liberal, -corporatist or -statist forms) which reflected the rise to dominance of the priorities of money capital combined with a general offensive against the working-class gains of the postwar boom. And, in the semiperiphery, the NICs continued to experience rapid growth, while elsewhere in the Third World the debt crisis severely constrained economic development.

By the 1970s, in a few semiperipheral states, the conjunction of autonomous local capital, relatively abundant urban middle classes and a nascent working class created the opportunity for rapid manufacturing growth.[23] This involved a locally based pattern of accumulation within the NICs and a specific linkage of these processes to the core, wherein the Department I of the core equipped a peripheral Department II, which then produced for the core markets of high but slowly rising incomes along with the peripheral markets with rapidly rising incomes. As with the monetary assets of the OPEC states, the emergence of the NICs produced a split within the periphery and a restructured interstate hierarchy. Commodity flows within the South in some respects paralleled those of the old North–South flows, and during the 1970s the rate of growth of manufacturing output in the NICs outstripped that of the core and the Eastern bloc. In general, however, the Third World experienced increasing differentiation.[24] The East Asian NICs have certainly received expanded levels of investment, primarily oriented to financial services and production destined for core markets. But elsewhere the picture has been quite different. Latin American production, whether by local or foreign capital, has been primarily for the potentially huge domestic markets. The fortunes of the OPEC states have fluctuated with the vagaries of the oil market over which they have only partial control. And finally, 'at least 75–80 developing countries have been shunted off to a side spur'[25] of low or even negative growth.

Much of the accumulation which did take place was financed by foreign capital – uninvested surpluses from the core, the OPEC surpluses and the monetization of debts – which was repaid by export earnings and the necessary recycling of loan capital to the core for the purchase of Department I goods. Indeed, it was this finance which was dominated by international private banking (based on the Euromarkets), and which created the international debt economy. The genesis and persistence of this pattern of accumulation depended on the continued expansion of the core markets (increasingly through credit expansion and fiscal laxity as the eighties wore on), while the rapid expansion of the Euromarkets combined with a lax monetary policy in the United States made credit easily available.

In addition to adverse trends in the semiperiphery itself, the response of the core to the 'second oil shock' was qualitatively different from that made to the earlier price rises. Carter began his administration in a similar manner to Nixon, attempting to reduce the value of the dollar and so encourage export growth, but he responded to the run on the dollar in 1977 by raising interest rates. The expansionary policy pursued in 1978, combined with the fall of the Shah and the continued expansion of the Euromarkets, resulted in periodic attacks on the dollar which were only placated by Volcker's monetarist turn in October 1979.[26]

This was dramatically illustrated by the movement of the three month rate of interest on the Eurodollar (deflated by the export prices of all the developing countries, including OPEC): it was 0 per cent in 1970–2, −30 per cent in 1974, around 0 per cent during 1975–8, falling to −10 per cent in 1979, before rising to an astronomical 20 per cent in 1981–2. It is in this sense that the difference between the first and second oil price rises was that the latter 'was "managed" through a contraction, and not just a shift, in world effective demand'.[27]

Although the onset of the crisis was due to a series of coincident national or regional crises of overaccumulation, triggered by an international event (the oil shocks) and aggravated by the absence of an hegemonic regime, the dominant mechanisms and countertendencies have been operating transnationally. In this context, it is not surprising that since 1973 transnational corporation profits have held up compared to those of domestic capitals, and the transnational banks have likewise derived substantial profits from the overseas investment of their excess liquidity (itself a consequence of the slump in the core).[28]

At the level of policy management, the tight fiscal and monetary policies pursued, together with the drift towards protection and the absence of international coordination with respect to fiscal, monetary, trade and energy policies, reflected the growing unilateralism of US hegemony along with the inability of the leading imperialist powers to organize co-operative solutions. The latter resulted from the combination of national economic rivalries and the emergence of an international credit system beyond national control (though not US manipulation), producing not only violent fluctuations across the foreign exchanges but also a new international debt economy. The contractionary nature of the fiscal and monetary policy within the core may be measured by the fact that real money supply growth in the OECD bloc decreased from 7.0 per cent per year in the period 1965–73 to 1.8 per cent per year for 1974–82; and fiscal policy has been consistently deflationary in every year since 1974 except 1975 and 1978 (at least until the end of the recession in the early eighties), with the important exception of the United States.[29] These conditions of depressed aggregate demand contributed to lower than trend levels of output and so reduced the rate of investment significantly. Between the periods 1967–73 and 1973–80 the growth rates of private nonresidential fixed investment decreased from 4.3 to 2.0 per cent per year in the United States, from 8.2 to 1.3 per cent per year in the European Community and from 13.5 to 2.4 per cent per year in Japan; and moreover much of 'the investment which has taken place since 1973 has been predominantly defensive and rationalizing: cutting costs by introducing new techniques but not markedly increasing potential output'.[30]

A Crisis of Overaccumulation

Theoretically speaking, then, the postwar growth in the core was based on the integration of a mode of capital accumulation, resting on a constant reorganization of the labour process which generated a steady rise in labour productivity and the per capita volume of fixed capital (technical composition of capital), with a continual adjustment of the norms of mass consumption to these productivity increases. Under such conditions, profitable accumulation depended upon increased productivity in Department I (capital goods) to offset the rising technical composition of capital, otherwise the proportion of fixed to circulating capital would become dangerously high and reduce the mass of profit; and the increase of productivity in Department II (consumer goods) had to match the rise in consumption norms, or else the share of wages in total income would expand at the expense of the profit share.[31] According to the French regulation school, these proportionalities were secured in the Fordist regimes of accumulation, which characterized growth in the core, through the construction of a complementary set of stabilizing institutions – a mode of regulation.[32] Typically, this involved class relations regularized through productivity bargains and the spreading of mechanisms of income maintenance to the whole population. In the leading, monopoly sectors of capital firms pursued oligopolistic price leadership in conjunction with accelerated depreciation. And the government sought to coordinate consumption and production through removing restrictions on the growth of consumer credit together with policies of demand management.

From this viewpoint, the crisis was a result of the barriers to the production of surplus value which derived from the exhaustion of old, Fordist regimes of accumulation and modes of regulation; in particular, limits to the growth of consumption were imposed by the saturation of consumer markets. A cognate argument has also been advanced by O'Connor in *Accumulation Crisis*.[33] The latter results from the conjunction of working-class opposition to capital (refracted through the capitalist ideology of possessive individualism, the money and commodity forms) and action through the state (often in tandem with nonclass consumer groups). Combined with rising proportions of fixed capital and the overproduction of credit (stimulated to stave off realization crises), the overall result has been the emergence of imbalances between Departments I and II, increases in the costs of reproducing labour power, and the undermining of the commodity form as rigidities were built into the social structures of accumulation.

Both of these approaches criticize general equilibrium theory – which underlies the analysis of, say, the McCracken Committee Report – for assuming that market mechanisms can ensure the necessary proportionalities between the branches of production and between production and consumption. Instead they argue that a specific framework of institutions is necessary to regulate these relationships, to provide the basis for a period of stable capital accumulation. But the general form of capital accumulation – that capitals have a competitive incentive to develop the productive forces without regard to the limit of the market, thereby generalizing production norms only through the devaluation of existing fixed capital – means that these conditions cannot be guaranteed. It is this uneven development of the productive forces within branches of production, the very means by which capital accumulation proceeds, which is simply ignored in these regulationist arguments. It is for these reasons that Clarke argued that 'the concept of the regime of accumulation [is] inadequate in stressing the systematic, as against the contradictory, integration of accumulation, and in stressing the discontinuities, as against the underlying continuity, of the various phases in the accumulation of capital'.[34]

The consequences of this general overaccumulation, and its associated uneven development of industries, sectors and countries, have been described above. The precise form of the crisis in the process of accumulation and the imbalances in the circuit of capital varied on a national basis and were determined by a variety of historically determinate factors, especially the intensity and scope of class struggle along with the institutional factors highlighted by the regulation school. In each case, despite continued recomposition of the social structures of accumulation, together with the profound reshaping of the global division of labour, the basis for a renewed 'long wave' of capital accumulation has not yet been laid. Moreover, throughout the 1980s, stimulating purchasing power through fiscal and monetary policies, in the belief that the problems were fundamentally those of realization, ran the danger of merely sustaining overaccumulation, thereby delaying the devaluation of capital necessary for renewed growth.

Thus, the continuing crisis of the capitalist world economy has been the result of the combination of the crisis of overaccumulation in the core, the uneven development of capital on a global scale – particularly the relative decline of US material preponderance and the consequent reorientation of US hegemony toward an increasingly predatory form

– and the associated monetary, trade and energy instability. Drawing upon and extending the structural trends in the postwar economy, the current restructuring is taking advantage of developments in telecommunications and data processing, the setting up of regional offices, the lightening of commodities and the lowering of transport costs, the rise of an international capital market, and production processes which are capable of disaggregation and that assign less significance to the skills of a national labour force.[35] All of these have allowed the production process itself to become internationalized. But this is not all. For regional – US, Japanese and Western European – competition over trade shares remains a central feature of the world economy despite the internationalization of production. This creates acute dilemmas for states, for as Block has pointed out two of the key pressures deriving from international economic relations are the competition for trade shares and the impact of international (money and productive) capital movements.[36] And, given the necessarily uneven accumulation of capital on a global scale, the requirements of augmenting trade shares will often contradict those of freeing restrictions on capital movements in search of the highest available rate of return. This is now a phenomenon which even the most powerful market, the US economy, has to contend with.[37]

Rising US Unilateralism or Trilateralism?

In the context of a general slackening of the pace of capital accumulation in ACCs, by the end of the 1960s it was apparent that the relative competitive advantage of the United States was declining with respect to the high growth economies of Northern Europe and Japan – this could be seen in growth rates, investment levels, productivity growth, and shares of world trade. In the monetary sphere, as early as 1961 the value of overseas dollar holdings exceeded that of US gold and foreign exchange reserves. The resulting Triffin paradox has been summarized by Brett as follows: while the *supply* of dollars to the rest of the world depended on the US balance of payments deficit, the *stability* of the dollar rested upon the medium- to long-term return of the US foreign balance to surplus.[38] Yet if the US economy was to return to surplus, and if the expansion of world trade was simultaneously to be maintained, a new form of international payments would have to be found: and the only possible sources for such liquidity were the new surplus nations, especially West Germany and Japan.

Meanwhile, US interests benefited in a variety of ways from the existing situation: the military–industrial complex and its attendant political representatives gained resoures from the high level of defence expenditure (this thrust was given renewed vigour by the escalation of the invasion of Vietnam from the mid-sixties); overseas dollar holdings could be used as credit in the foreign expansion of US transnational corporations; and finally, 'the American balance of payments deficit was an outflow of paper money in exchange for which the USA obtained some very concrete returns'.[39] On the other hand, the surplus countries faced a cruel dilemma: if the deficit escalated the United States might devalue, thus leading to a loss of reserves; however, the growth of domestic surpluses could feed through into inflation, thereby undermining international competitiveness. The inability of either side to impose the necessary adjustment costs upon the other, or their reluctance to share such costs, resulted in a stalemate.

In August 1971 the deadlock was terminated by Nixon's unilateral ending of dollar–gold convertibility, refusal to guarantee the value of the dollar and imposition of temporary import controls. These events clearly marked a turning point in the management of US hegemony over the international financial system, signalling a shift to the position of a predatory hegemony. For Nixon's actions amounted to the unilateral abrogation of the 'Atlantic constraint'[40] by Washington; the break with gold and the subsequent collapse of the fixed exchange-rate regime, as Parboni notes, 'eliminated any need for the United States to control its own balance of payments deficit, no matter what its source, because it was now possible to release unlimited quantities of nonconvertible dollars into international circulation'.[41] The immediate effect of the dollar devaluation was a US export offensive against West European and Japanese market shares, particularly in agricultural produce and military supplies. The successful export drive of 1974–5 depended upon a combination of a domestic recession (to depress oil imports) together with the temporary advantage of the 1971–3 devaluations. By these means, US competitiveness was somewhat restored in foreign and domestic markets. However, a more far-reaching consequence was the manner in which the 1971–3 devaluations facilitated the US economy's adjustment to the oil price rises of 1973–4. As suspected in some quarters at the time, these rises were to a degree encouraged by the Nixon administration, as they would help the US in a number of ways: high-cost wells restored to profitability, increased funds flowing to allies in the Middle East, rapid improvements in the profitability of the majors, improvements to the balance of payments (through arms sales and dollar recycling into US assets), and a competitive blow dealt against Western Europe and Japan.

Meanwhile, a deadlocked Congress refused all legislative attempts to reduce domestic oil consumption. And so, before long, the US response to the oil price rises shifted from real adjustments (increased exports and/or reduced consumption) to what Calleo appropriately terms 'monetary manipulation'.[42] For in the second half of the seventies domestic inflation reduced the real cost of US imports, while the continued dollar depreciation against the mark and the yen reduced the price of crude outside the dollar zone. Oil producers thus found their income squeezed by inflation on the one hand, and through the devaluation of both their purchasing power (the bulk of Middle East imports came from Europe) and their assets (which were largely denominated in dollars and sterling – currencies that moved together) on the other. OPEC was relatively powerless to prevent these developments as a result of the temporary oil glut, intra-OPEC disputes and US foreign policy towards Iran and Saudi Arabia (in particular, massive arms sales).

It was against this background that Kissinger's 'Year of Europe', announced in April 1973, should be understood. As he explained:

> [The United States] had global interests and responsibilities . . . [while] our European allies have regional interests. [And whereas] in economic relations, the European Community has increasingly stressed its regional personality . . . the United States . . . must act as a part of, and be responsible for, a wider international trade and monetary system. [The task ahead, therefore, was to] reconcile these two perspectives.[43]

One aspect of this reconciliation was to be the manipulation of the 'inordinate privilege' of the dollar described above. Throughout the 1970s the international monetary disorders thus produced were exacerbated by the oil price rises of 1973 and 1979, which had the effect of transforming the OPEC states into surplus countries, while seriously impoverishing many Third World oil importers and temporarily throwing the OECD economies into balance of payments difficulties. Once again, the problem of adjustment costs, now overdetermined by a profound restructuring of the international circuits of productive, money and commodity capital, posed acute problems. The surplus countries were unwilling to back a new form of international liquidity unless the United States took steps to reduce its deficit through deflationary policies at home, and the United States wanted the surplus nations both to adopt expansionary policies (to suck in its exports) and to bear a greater percentage of the Atlantic alliance's unproductive military burden. In turn, the surplus states rejected these policies as inflationary and therefore liable to undermine international competitiveness at a time of general stagflation and mounting unemployment.[44]

This turn towards US unilateralism gainsaid the putative offer of a nascent *trilateralist* strategy to share economic leadership, and it was complemented in the geopolitical sphere by a shift in US–Soviet relations. Already in the early 1960s – and in response to the failure of the European Defence Community plan, the Suez crisis, the Soviet successes in space and in the Third World – Kennedy had advanced a strategy of 'slowly undercutting the foundations of the Soviet order' through Atlantic unity. However, the Kennedy drive for Atlantic partnership ran adrift on France's veto of British EC membership.[45] And not only did US initiatives towards Western Europe falter – Brandt's *Ostpolitik* instead of Kennedy's unity – but the United States also assumed those geopolitical roles relinquished by Britain and France between 1956 and 1965. Together with the escalation of the US invasion of Vietnam, this created a space in the Third World for both Soviet intervention and a series of West European economic ventures. Within the Eastern bloc, despite advances towards rough parity in the field of strategic nuclear weapons, the Soviet Union faced major difficulties.[46] Economic problems of low labour productivity, combined with a growing technological gap with the West, dictated – under the pressure of the need to raise living standards – a strategy of high-technology imports from the West. Moscow's control over Western communist parties had declined as a result of the emergence of Eurocommunism, while the split with China in 1963 faced the Soviets with two fronts. And finally, the costs of policing Eastern Europe (most brutally exposed in Hungary in 1956 and Czechoslovakia in 1968) began to rise from the mid-sixties. In consequence, the Soviet Union had a clear interest both in reaching agreements with the United States on economic trade and in normalizing the geopolitical balance in Europe.

In sum, by the late 1960s, the United States no longer had a decisive and unambiguous military superiority over the Soviet Union, it did not have unrivalled leadership over the capitalist West, and it found increasing difficulty in ordering the political and economic systems of the Third World. By 1967–8 the US elite had experienced not only currency and gold crises, but also the Tet offensive in Vietnam and worker–student unrest domestically and in Western Europe.[47] It was the conjunction of these developments with the economic problems noted above that was to lead to the crisis of Atlantic integration of the 1970s and beyond. Between the superpowers, the upshot of all this was that the next decade (1969–79) marked a halting period of East–West *détente*, as the United States and the Soviet Union sought to detach their mutual rivalries from their intrabloc concerns and to pursue negotiated arms control. Meantime, however, the United States attempted to share the costs of containment by devolving regional policing to its allies (the Nixon Doctrine), while playing the China card against the Soviet Union.

The Oil Order in Transition: Oil Capital, OPEC and the United States

The structure of what Bina has termed the 'traditional' argument in relation to the origins of the price rises refers 'to (1) the primacy of demand and supply in regulating market conditions, (2) the conventional oligopolistic theory of the firm, and (3) the traditional view of *cartel* associated with OPEC'.[48] This case typically regards OPEC as the prime mover in the events of the 1970s, representing an assertion of the power of producer-states against the (predominantly US) companies of the West, the consuming nations and the hegemony of the United States in the international system. And what made these OPEC actions possible was the relative decline of the share of US reserves and production. As we saw in chapter 3, however, by the late 1960s the US-ordered world oil industry faced a number of closely connected challenges. The decline of reserves and production to consumption ratios within the United States did result in the country becoming a net importer of oil, with adverse consequences for the balance of payments. But in the industry itself, the expansion of the independents and of the West European and producer-state national oil companies undermined both the degree of concentration and the opportunities for collusion. Combined with the altered overall position of the US market, these cracks in the majors' oligopoly changed the power relations between the producer-states, the companies and the consuming nations. This increased competition resulted, further, in a loss of price control: as the inherent tendency to overproduction reasserted itself, so the level of surplus profit in the industry fell and the struggle over its distribution intensified. Finally, the growing struggle over the distribution of the rent coincided with larger demands on behalf of each of the recipient parties.

Of equal importance was the simultaneous series of challenges to US hegemony which became increasingly apparent, and among which a declining share of the world's oil production was only one aspect. First, the conjunction of the US defeat in Vietnam and the Soviet approach to something like strategic parity in the nuclear arms race, together with the latter's expanded naval and airlift capabilities, meant that the

US capacity for regional intervention was drastically curtailed, while the scope for Soviet support to Third World revolutions or to radical nationalist regimes increased. Second, the rate of growth of the US economy faltered in relation to that of Japan and Western Europe, with the inevitable result that the hegemony of the dollar was increasingly if fitfully challenged. Third, with the advent of *détente* and the consequent reduction of US tutelage *vis-à-vis* its allies, West European states and Japan began to develop diverging interests throughout the Third World, particularly in the Middle East, Fourth, the general economic crisis exacerbated both domestic conflicts and international tensions. In this broader context, therefore, the Nixon–Kissinger team was keen to find an issue around which the United States could regain its undisputed role as leader of the West. And events in the oil industry were to provide just such an opportunity – or so they reckoned.

A Changing Industry

During the 1950s the majors had accepted a rise in the payments to the producer states because these were deemed by the US Treasury to be an 'income tax' which could be offset against domestic tax liabilities (rather than royalties which would be treated as a cost of production). This made it more profitable for the majors (and later the independents) to attribute the greater proportion of their earnings to the production phase, thereby reducing their declared downstream earnings. The consequence of this was that the companies could agree to increases in producer-state revenues, so long as these were in the form of rising fiscal charges associated with an equivalent rise in the posted price for crude oil.[49] For successive US administrations this 'golden gimmick' admirably served their foreign policy aims. As a US Senate committee put it:

> The foreign tax credit was an instrument of US foreign policy. US foreign policy objectives were threefold. First, the US desired to provide a steady supply of oil to Europe and Japan at reasonable prices. . . . Second, the US desired to maintain stable governments in the non-communist pro-Western, oil exporting countries. Third, the US desired that American-based firms be a dominant force in world oil trade. These three . . . goals were generally attained during the 1950s and 1960s.[50]

Now, although the share of the price of crude accruing to the producer states increased throughout this period, it was more than offset by the worsening terms of trade with the West's manufactured goods.

And, at the same time, the incursion of the independents into the oil arena had weakened the bargaining power of the majors with respect to the producer-states. (An important example of this changed balance of forces can be seen in the events in Algeria and Libya in 1970–1.) Therefore, beginning in the early 1970s, the producer-states pushed for increased rent from and control over their oil production. This was true of such radical regimes as Algeria, Libya and Iraq as well as the conservative states (Saudi Arabia and Iran) desiring large-scale arms transfers.

The majors resisted, however, because despite their advantageous tax arrangements they had seen their share of the surplus profits squeezed between the twin forces of a declining rate of surplus profit (due to a fall in the real price of oil) and the increased rents of the producer-states. Still the majors' control over the circuit of oil capital was declining. Blair's research uncovered some aspects of the response to this predicament, which included an attempt both to reduce reliance on crude production for profit generation and to improve the ability to attract external capital.[51] To this end, and in order to open the domestic market and thus relieve overproduction elsewhere, the majors tried to gain control over the US independents. At the same time, an attempt was made to widen the profit margin in refining and to raise the target rate of return on investment: between 1959 and 1966 profit rates for US direct foreign investment in petroleum dropped from 30 to 10.8 per cent before rising to 34.4 per cent in 1972 and over 30 per cent again in 1974, falling to about 18 per cent during the second half of the seventies and rising to 37 per cent in 1979.[52] (The scale of this investment is indicated by the fact that in 1972, 49 per cent of US foreign profits came from petroleum companies.)[53]

These historically low profit rates of the early to mid-1960s came at an unfortunate time. In the first instance, during the 1950s and early sixties the majors expanded into the currently highly profitable petrochemical business. However, because of the long lead times in such projects, this new capacity came on stream just as long boom faltered and demand fell. A second wave of diversification began in the early 1960s when the majors entered the competitive fuel industries, particularly coal and nuclear power. In addition, the majors were contemplating production in new high-cost areas, especially in the North Sea and Alaska. Traditionally, the oil industry had been self-financing to a high degree, yet such profit rates were no longer sufficient for this to be possible given the planned levels of investment. Indeed, until the mid-1960s oil capital raised some 80 per cent of its finance internally (90 per cent in the case of the majors) whereas the percentage of funds

coming from external sources for the foreign operations of US petroleum companies over the period 1966–72 increased to an average of
51 per cent.[54] Meantime, the 1960s were an era when the oil companies'
projections for future demand and production levels were still extremely
optimistic, and by 1970 or thereabouts the majors had a clear interest
in reorganizing the industry or in price rises to increase their profits.[55]
Finally, as Nore has suggested, it is possible that a more defensive
corporate strategy was at work in so far as the majors could foresee
that, in order to maintain their vertical balance and provide outlets for
their technological expertise, they would have to opt out of direct
ownership in some areas altogether.[56]

The US administration also had clear interests in an increased price
for crude, and this was clearly – if not diplomatically – signalled to the
OPEC members during 1972 and 1973.[57] Such rises would help with
the problems of rising domestic production costs and so make many
wells viable again, thus improving (at least in the short term) the degree
of national self-sufficiency along with the balance of payments position.
Bina has shown that the decade from the late 1960s to the early 1970s
saw US per barrel exploration costs rise by 35 per cent and development
costs by 195 per cent.[58] A price rise would also continue to lubricate
US foreign policy in the Middle East, as the oil-backed regimes (in
Saudi Arabia and Iran) would be strengthened – that was the calcu-
lation: the State Department did not know that rent-based states tend
to have weak social bases – and any recycled petrodollars could buy
the military hardware the Nixon Doctrine required for regional clients.
Obviously, price increases would directly benefit the US majors through
a revaluation of their assets and an immediate improvement in their
profit positions, a concern of immediate interest to the State Department because of its close company linkages. Finally, a price increase
would deal a competitive blow to the United States' major trading
competitors in Western Europe and Japan (because of their greater
import dependence on oil and the pricing of it in dollars).[59]

It is in this context that the struggles over prices and control should
be seen. What began as a struggle between the majors, independents
and producer-states over the distribution of the surplus profit – centred
around the level of the posted price – was transformed into a structurally
determined coincidence of interests between the oil companies, producer-states and above all the United States for an increase in total surplus
profits generated in the oil sector. Central to this process was the
changed strategy of the US state towards its regional clients in the
Middle East (itself a reflection of broader developments in East–West
relations) along with its desire to reconsolidate hegemony over the

capitalist West through trade advantage and monetary manipulation. Equally, however, the space opened up in such a strategy for a few determined radical regimes (above all Algeria and Libya) to pursue independent strategies was considerable. Unfortunately for the US administration, the unintended consequences of its actions thereby eluded the control of either the US state or the oil companies. For in the early 1970s the oil market strengthened because of the synchronized boom of the OECD bloc and the rapid increase in US oil imports, thus helping to transform the oil arena from a consumer to a producer market. These market developments were exacerbated by the coincidence of a shortage of refinery capacity in Western Europe and Japan, the closing of the Trans-Arabian pipeline and the Arab boycott during the 1973 Arab–Israeli War. With the United States and the majors no longer in control of either the circuit of oil capital or the strategies of the Middle East producer states, the eventual price increases were far larger than intended and the attendant dislocation to the global economy that much greater.

The Moment of OPEC

The large-scale changes began on 12 June 1970 when Libya put pressure on the independent company Occidental by ordering progressive cuts in the level of production. Occidental was solely dependent on Libyan crude and was refused assistance by Exxon – the subsequent and inevitable acceptance by Occidental of Libya's conditions marked OPEC's first 'victory'. Events moved quickly, culminating in a wave of nationalizations and price increases of 1973–4 (see below). As a consequence, the majors' control over the productions of non-Communist oil fell from 82 per cent in 1963 to 32 per cent by 1974; contrariwise the control of the national oil companies (NOCs) of the producer-states increased from 9 to 61 per cent.[60] Prior to these changes the majors earned the bulk of their returns upstream, whereas after 1973–4 their profit structure shifted downstream. Indeed, there is some circumstantial evidence to suggest that companies moved their profits downstream during the period 1971–3 in anticipation of the forthcoming changes.[61]

In some contrast to this loss of control over production, the Majors' control over refinery capacity only fell from 65 per cent in 1963 to 50 per cent in 1974 (this is because the bulk of the world's refinery capacity was and is located in the consumer-nations) and similarly their share of marketing capacity only declined from 63 to 45 per cent.[62] Indeed, Luciani has shown that the vertical integration of the largest eighteen oil companies actually improved as a result of the changes in the 1970s, at least as compared with the late 1960s. By 1980 they only controlled 38 per cent of the crude they had in 1971 and this represented about half of their needs, but including the crude available on long-term contracts the companies still controlled 76 per cent of their previous total. Because in 1971 they were crude-long, this constituted an improvement on their former position of vertical integration.[63]

Less pleasing was the fact that in contrast to the previous, interterritorial planning of upstream investment, taking account of the downstream needs of the majors, the NOCs now planned investment on a national basis, with a reduced control over downstream activities. The closed, state-supported and collusive trading circuits of the old order were supplanted by a much more complex cartography of oil trade. This now involved bilateral contracts, barter deals, world trading and spot markets. The NOCs have found themselves crude-long and the majors have been crude-short. To cope with this, directly marketed crude rose from 8 per cent of the total in 1973 to 45 per cent in 1980. Initially the NOCs were obliged to leave marketing to the majors: in 1975 the latter controlled 25 million barrels per day compared with a volume refined of 19 million barrels per day. But the earlier proliferation of independent refiners and temporary shortages led the NOCS to enter marketing directly; and, furthermore, the consumer-nations moved into direct trading for reasons of security, such that by 1980 the majors were buying 7 per cent of their crude on the market.

The result of these changes, as Levy has pointed out, is that the oil market is now two-tiered.[64] The production in the old areas is now dominated by producer-state NOCs but elsewhere (non-OPEC and new areas) the majors are still strong. The share of traded crude taken by the majors has fallen dramatically. The second tier consists of deals between the producer-state NOCs and a wide diversity of buyers, including states, Comecon, independents, and traders. Throughout these developments, the OPEC states have continued to differ in terms of their abilities to absorb rents, their reserves and their political structures. They have also faced constant problems of price differentials, marketing control, conservation and the lack of harmony with regard to their upstream investment, features which have been compounded by OPEC's declining market share in conditions of falling (slowly rising from 1984) total demand. Overall, these events have resulted in, on the one hand, a breakup of the old regime and the rise of NOCs, and on the other, a geographically diverse pattern of exploration and production.

In these circumstances, the majors initially appeared to be moving towards becoming more broadly based energy and mineral corporations and there was a clear trend in this direction, presumably financed by the windfall profits derived from the price rises. As the companies lost control over production they lost the advantages of vertical integration, while the changed circumstances dictated the development of new investment strategies. The Italian state energy concern ENI has researched these new strategies, concluding that the majors have adopted a threefold strategy.[65] In the first place, they have sought to acquire new reserves in politically safe areas, especially in the United States through takeovers and mergers. This then involves, secondly, the necessary infrastructural investment for the development of these new reserves. And, finally, this demands the reequilibration their own vertical integration (rather than the horizontal diversification many predicted), as well as strategies to ensure that the necessary capital is forthcoming.

Price movements have both prompted and facilitated this restructuring. Between 1970 and 1974 the nominal price of a barrel of oil rose from $1.8 to $10.46; in real terms ($1976) the rise was from $2.85 to $10.46; and during 1973–4 the OPEC producer-state revenues expanded by some $64 billion – equivalent to 1.5 per cent of world capitalist output.[66] As a result of the recession, demand for oil fell, albeit with important regional variations, more or less continuously between 1979 and 1983. This was the result of four interrelated factors. First, the price rises produced a search for substitutes and the discontinuation of oil-intensive processes; second, the international recession in itself reduced demand; third, changing patterns of economic activity – primarily a shift from manufacturing to services – resulted in a lower total energy requirement per unit of gross domestic product; and fourth, the consumer-nations promoted policies to reduce oil/energy consumption (such as taxes and energy conservation programmes).

Overall, the exact levels of surplus profits going to either party at the upstream end of the industry varied with the terms of the joint venture agreements and the service contracts negotiated. In many cases, the majors have lost control over production to the NOCs of the producer-states. The degree of control – through joint ventures, service contracts, production sharing agreements, buy-back deals, and other means – remaining with the companies varies but, as Luciani has shown, is by no means as diminished as many suppose. The arrival of state monopolistic NOCs at the upstream phase, together with the continued reliance of these companies on the majors for technology or marketing, makes generalization about the industry difficult.

Oil and the Economic Crisis

There can be little doubt that the oil price rises were important in determining the timing and conjunctural form of the international recessions during the 1970s. Equally, however, the long boom was facing the severe difficulties discussed above and there is no reason to suppose that a nonrecessionary, crisis-free resolution was possible.[67] Moreover, the price increases themselves were not the result of an arbitrary act of Arab 'nationalism' (the epithets in the Western media were frequently racist and occasionally militarist), but were intimately bound up with the broad changes in the capitalist world economy and the interstate system. Specifically, the changing pattern of US hegemony played a crucial role in mediating these events.

Theoretically speaking, it is possible to identify a number of ways in which the price increases of oil and non-oil primary commodities contributed to the economic stagflation of the OECD bloc during the 1970s. First, the price rises worsened the terms of trade, especially for manufacturing industries, and this combined with a lack of compensating increases in productivity growth and real wage resistance produced declining profits and hence investment. In turn, the decline in investment demand led to a fall in output, productivity growth and dynamic efficiency. Second, the immediate balance of payments disequilibria were matched by deflationary fiscal policies which also reduced aggregate demand and hence output. Third, the rise of consumer and input prices contributed to the adoption of tight monetary policies which had similar consequences. And fourth, the rising consumer prices, combined with real wage resistance, also resulted in declining profitability with its attendant effects. These mechanisms are summarized in table 4.1.

Let us now consider some of the empirical evidence relating to these issues. Perhaps the highest estimate for the effect of commodity price increases is due to Bruno who has argued that the direct effects of the primary commodity price shocks of 1972–4 can explain some 40–60 per cent of the slowdown in the rate of manufacturing growth.[68] His evidence for this is drawn from a series of regression analyses that indicate a correlation, on a country-by-country basis, between productivity slowdown and movements in input prices. But, as Bruno himself notes, much of the cross-national movement in input prices is a result of real exchange rate movements, and as Jackman points out these must be considered to be at least in part endogenous.[69] Therefore, Bruno's figures are likely to be a considerable overestimation, especially given

TABLE 4.1   Some economic consequences of a rise in commodity prices

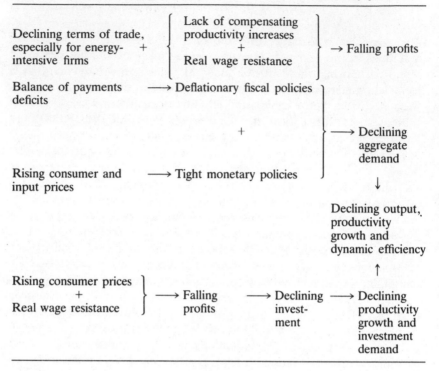

the increasing importance of capital (rather than trade) movements to exchange rate determination during this period.

For oil alone, Nordhaus concluded that the first oil shock had only a marginal impact on the OECD economies until 1979: it could account for 6 per cent of the decline in gross national product growth rates, 11 per cent of the increase in inflation, 10 per cent of the increase in unemployment, and 6 per cent of the decline in the rate of productivity growth.[70] Moreover, these increases cannot be considered as exogenous for the reasons argued above. And, in addition to the long-term changes noted above and the contingent political, speculative and climactic events, the worsening terms of trade between Western manufacturers and the primary producers of the Third World were caused by a slower rate of capital investment in the primary producing areas than in the manufacturing sectors during the long boom.[71] In other words, these 'shocks' were not the result of an exogenous disturbance to an otherwise harmonious market order but the direct result of the necessarily uneven development of capitalism on a global scale, together with the related periodic restructuring of the interstate hierarchy.

None the less, despite these caveats, the worsening terms of trade – 50 per cent for manufactures against all primary commodities between 1972 and 1974, improving from 1975 to 1978 because of inflation and a slackening oil market before worsening again by 40 per cent as a result of the second oil price rise – did reduce growth in the OECD bloc, and with respect to oil this was particularly so for energy intensive sectors.[72] The OPEC states ran a total balance of payments surplus that averaged $40 billion per year over the period 1974–8, with the surplus declining from $70 to $15 billion.[73] The deflationary impact of these OPEC surpluses was due not so much to its direct effects – the 'OPEC tax' – as to the financial problems entailed. Given the oil deficits of the OECD countries, and their short- to medium-term inability to compensate by reduced consumption or increased exports, deflationary policies were adopted to conserve reserves. In one sense, of course, the financing of these deficits was automatic (this was not true for the less developed countries (LDCs)) in so far as OPEC states had little immediate choice but to invest their surpluses somewhere in the OECD bloc. The problem of recycling was not so simple. Indeed, Tew has argued that there is little evidence that exchange rate movements played a significant role in equilibrating the international payments system as between different countries with their varying shares of deficits and OPEC investments.[74]

The impact of the deficits was highly uneven as between OECD nations and, more importantly, between the ACCs and the LDCs. Similarly, the ability of different national economies to respond to the deficits was highly variable. Japan, Germany and the United States (at least until 1976) were able rapidly to absorb their deficits through compensating export drives. On the other hand, the non-oil LDCs increasingly turned to the international banking system for finance and many such countries were plunged into serious deficit. The ensuing polarization of surpluses and deficits spread the international debt economy world-wide and posed a vicious obstacle to economic recovery.[75] The rise in oil prices also eliminated the 'dollar overhang' which had developed – for the price increases contributed to the rapid rise in the dollar value of world imports, while the OPEC countries generated large surpluses that were recycled to pay for the West's increased import bills.[76] After 1974 the international financial disequilibrium was being financed rather than cured, such that the second price increase of 1979 meant that the disequilibrium was back to square one. Moreover, the monetarist reaction to the second oil shock in the core was qualitatively different to the recycling of the first period and eventuated in a contraction and not just a shifting of world aggregate demand, although this reaction itself was not primarily driven by the problems caused by the oil price rises.[77]

There can be little doubt, then, that the oil and commodity price rises were a contributory factor to the nature and timing of the onset of stagflation, but less often recognized is the limited role that oil played in the end of the postwar boom. The oil and commodity price rises mediated the overall structure of the crisis and were a proximate cause of some of its most visible features – financial and monetary instability – but they were not the determinants, or even principal features, of the crisis. They are best seen as symptoms. For the uneven global development of capitalism implies that there is no guarantee that the rate of growth of primary products is sufficient to meet the warranted increase due to the growth of production in the secondary and tertiary sectors. Disproportionalities of this kind are inevitable. In this context, the inflation of 1968–71 and beyond was exacerbated by the commodity price increases.The rise of commodity prices contributed to inflation directly in so far as producers sustained profit margins through mark-up pricing and indirectly through real wage resistance. But the persistence of inflationary trends throughout the recessions showed important cross-national variations which had little to do with developments in the primary sector, let alone the oil market, and everything to do with the institutional form of capital accumulation, class relations and the state system of the countries concerned.

OPEC, Saudi Arabia and US Hegemony

Introduction

As noted in chapter 2, realists and transnationalists agree that hegemony rests primarily on material preponderance. And the case of oil illustrates this argument all too clearly, for whereas the United States produced 62 per cent of the world's oil just prior to the Second World War, in 1972 it produced only 21 per cent.[78] Krasner has drawn the obvious conclusion, and it is one which neatly characterizes the consensual interpretation:

> The decline of American power and changes in postwar economic regime are nowhere more dramatically evident than in the area of crude petroleum. . . . The bargaining power of the major corporations declined as a result of increased host-country knowledge of technology and the market, and the growth of independent companies. The power of host countries

increased vis-a-vis consuming countires because of the exhaustion of surplus in non-OPEC producing countries, particularly the United States.[79]

In addition to the neoclassical economic arguments mustered here, realism serves as the guide to the political consequences of these changes. For if the power of a state is determined by the resources within its territory, then the relative decline of US reserves, along with the assertion of national control over the previously foreign-controlled resources of the Middle East by the indigenous states, must amount to a fundamental shift in the distribution of power in the world system. More specifically, if hegemony rests upon material preponderance (realism), combined perhaps with the ability to order international regimes (transnationalism), then the apparent loss of US dominance and control signalled by the OPEC events must indicate a significant erosion of hegemony. The economic case underlying this argument has been criticized above, but the implications of these developments for the fortunes of US hegemony also need to be addressed.

Parboni has noted that the postwar US balance of payments position went through three states.[80] In the fifties the trade surplus was used to finance an outflow of aid and military spending; the sixties saw the outflows taking the form of direct investment (especially to Western Europe and Canada); and the seventies – marked by the ending of dollar–gold convertibility – witnessed US deficits due to rising net oil imports, rising deficits in capital flows for direct investment and a shift in banking funds offshore. The oil component of this deteriorating position was to become increasingly significant. In the 1950s and sixties the United States still had sizeable spare capacity and was able to cope with the disruption to Western oil supplies due to the Suez crisis of 1956–7 and the 1967 Arab–Israeli War.[81] But import controls hastened the time when the United States became dependent on oil imports and this had deleterious consequences for the balance of trade. Price controls in the United States during the early 1970s, culminating in the introduction of 'two-tier pricing' in August 1973, together with the final abolition of import quotas, produced an upward leap in imports. Thus, during the seventies the United States moved from being 90 per cent energy self-sufficient to importing 50 per cent of her requirements – including some 30 per cent of OPEC output – thereby consolidating OPEC's market strength, with the market share of the latter peaking in 1973 at 55.5 per cent.[82]

Stimulated by this increasing vulnerability, the United States sought both to encourage non-OPEC sources of oil production (in particular the North Sea) and to establish stronger client–patron relations with Saudi Arabia and Iran. With regard to US–OPEC trade, the United States supplied some 15 per cent of OPEC's foreign purchases yet still its trade deficit in the latter half of the seventies was largely accounted for by its deficits with OPEC.[83] But the balance of *payments* implications were rather different. In 1978 (before the second oil shock) only 23 per cent of the total energy requirements of the United States were met by oil imports, whereas this percentage was 92 per cent for Japan and 67 per cent for Germany. Furthermore, for the period 1974–7 the total purchases by the United States from OPEC were $106 billion but the purchases by OPEC from the United States were $70 billion and the known OPEC investment in the US (primarily in Treasury bonds and corporate securities) was a further $38 billion.[84] And so the dollar devaluations cannot be attributed in any simple manner to the oil price rises. Indeed, through the OPEC purchases of US assets and the recycling of oil income for military equipment, the US payments position may even have improved.

Inside OPEC itself, the 'cartel' has been unable to plan the overall production of crude (at least until 1983) and was limited to price setting and the coordination of its member states' activities with respect to the oil companies. In fact, during the 1960s Iran maintained a resolutely independent stance (especially after 1963), while Saudi Arabia systematically opposed every attempt to institute some form of prorationing until 1982. Indeed, the rivalry between these powers, especially after the British withdrawal from the Gulf, combined with the fact that Saudi production only became predominant in the early to mid-1970s, weakened the Kingdom's ability to dominate OPEC. To grasp this aspect of OPEC behaviour it is necessary to consider the central role of Saudi Arabia. After the events of 1973–4, the Kingdom was the third largest oil producer (following the United States and the Soviet Union), produced by far the largest share of traded oil (about one-third of the total) and it was (and is) the country with the largest reserves. In the 1970s the Kingdom's per capita proven reserves were some three times those of Iraq and ten times those of Iran, while the comparable position in Kuwait and the United Arab Emirates amounted to little, given that their joint production was less than half that of Saudi Arabia.[85] Based on this commanding position, the Saudi strategy developed in the late 1970s is generally recognized to revolve around three elements. As summarized by Roncaglia, these are as follows.[86] In the first instance, the Saudis tried to pursue a profitable strategy of high unit royalties at the expense of a reduction in their market share. To secure this aim, the second element was the desire to use their market control to avoid erratic supply and price fluctuations, while inducing a stable trend in oil prices – the unrequited 'Yamani-formula' of 1978 proposed to link crude oil prices to the rate of growth of nominal gross national product of the industrialized countries – over the medium to longer term. And finally, the Kingdom sought to maintain its position as the leader of OPEC by having a sufficient market share to be able to influence market prices and sufficient excess capacity to discipline recalcitrant members.

In relation to this composite strategy, the United States has been placed in a paradoxical position: it desired a reduction in the real price of oil (though not a return to former levels), but the oil–profit–arms recycling strategy, as well as the regional demands of the Nixon Doctrine, necessitated high prices. A number of developments conspired to transform the situation in the Middle East. For while the position of Washington was enhanced by Kissinger's mediation in the wake of the 1973 war, and while the Soviets had suffered a major set-back in Egypt (1972), the Arab states had both instituted an oil embargo and grown in power as the United States became more dependent on Middle East oil imports. On the other hand, although Syria and Saudi Arabia refused any participation in the Camp David process, no radical Arab coalition emerged. In fact, with the isolation of Syria and Iraq within the Arab world, alongside Saudi concern with peninsula and internal matters (King Faisal was assassinated in March 1975), the US mediation of the Israel–Egypt conflict strengthened the Arab moderates. Meantime, the Ford and Carter administrations massively increased arms sales to the region, such that 'by the mid-1970s, sales to Iran, Saudi Arabia, and Israel constituted over one-half of America's global Foreign Military Sales (FMS) program'.[87] Iran remained the favoured policeman in the 'twin pillar' policy, but sales to Saudi Arabia were increasingly aimed at influencing its role within OPEC. As Kupchan has pointed out:

> Though repeatedly denied by administration officials, it was clear by the second half of the 1970s that the United States was selling weaponry in order to influence oil production and pricing policy. By providing Saudi Arabia with the weapons they wanted, the United States hoped to convince the Saudis to maintain production levels that would ensure sufficient supplies asnd reasonable prices. . . . From 1973 to 1978, the annual figures for total Saudi arms imports were (in 1979 dollars): $124, $482, $323, $542, $1017, and $1194 million, respectively.[88]

It was also fortunate for the United States that the period 1975–9 constituted something of a lull in world oil markets despite the consolidation of upstream control by the producer states.

OPEC, 1959–79: From the Maadi Pact to the Yamani Strategy

In April 1959 the largest oil exporters, Venezuela, Saudi Arabia, Iraq, Kuwait and Iran, together with Egypt and Syria, over whose territory 65 per cent of Middle East crude travelled, signed the Maadi Pact.[89] The pact had limited aims centring on regular consultation among the members to co-ordinate policy towards the oil companies. By September 1960 the Baghdad Resolutions, which included additional proposals for production sharing and joint action in the face of hostility from the majors, formed the basis for the establishment of OPEC. At the same time the Soviet Union announced that it would limit its oil exports and avoid competition with OPEC members. The majors' response, no doubt emboldened by the assertion of the New York Times that OPEC constituted 'an interference with the principle of free enterprise', was to refuse to recognize OPEC and to exploit the tensions between the member states. And indeed, the early attempts to develop a programme of production sharing were a complete failure, undermined by the machinations of the companies and the self-interest of the member states. In fact, it was not until June 1968 that the organization was able to define a coherent 'Declaration of Oil Policy', one 'which was to serve as OPEC's guiding light during the following five years'.[90] In essence this document committed the member states to co-operate in the pursuit of participation, the renegotiation of concessions and producer-state control over pricing policy.

It is a significant and often overlooked fact that the first successful challenge to the majors' monopoly over pricing policy was made not by OPEC but by Algeria and Libya in alliance with Egypt, Iraq and Syria. Just prior to the opening of negotiations on price increases between Libya and the relevant oil companies (January 1970), Algerian, Libyan, Iraqi and Egyptian officials met to discuss policy. Meanwhile, in the summer of 1969 Iraq had signed an oil agreement with the Soviet Union, as did Libya in March 1970. And in May 1970, after a 'bulldozer belonging to the Syrian Ministry of Public Works "accidentally" smashed into a section of the Tapline',[91] the Syrians refused the Tapline's owners permission to repair it, thereby tightening the Mediterranean oil market. In these circumstances, and given that the majors and the United States wanted price rises in any case, the US State

Department refused to back those companies that wanted to stand firm. Encouraged by the success of Algeria and Libya, the twenty-first OPEC conference in December 1970 adopted strong and precise resolutions on pricing policy. At this juncture the majors sought and were granted exemption from US anti-trust legislation to negotiate with OPEC *en bloc*.

But under the prompting of James Akins of the US State Department, the relevant oil companies concluded a separate agreement with the Gulf producers in Tehran (February 1971). Undaunted 'Libyan officials immediately went on the offensive', while those OPEC members who also exported to the Mediterranean – Algeria, Saudi Arabia and Iraq – somewhat surprisingly 'decided to give Libya a mandate to negotiate for all of them'.[92] This move was not without profit as the Tripoli price increases (April 1971) were some 60 per cent higher than those gained in Tehran. All of this counted for little, however, when the United States abandoned the dollar's convertibility and devalued in August 1971; and at its twenty-fifth conference (September 1971), OPEC demanded that the Tehran–Tripoli agreements be renegotiated accordingly. This was only agreed after considerable delay by the majors, as was a further adjustment after another devaluation in 1972. By 1973 control over pricing policy had passed to OPEC, and during the 1973 Arab–Israeli War OPEC announced a major, unilateral price increase from $3.0 to $5.1 per barrel. And finally, the Arab oil embargo resulted in sharply rising spot prices which individual members took advantage of, while OPEC consolidated these gains by announcing a new posted price of $11.65 per barrel on 23 December 1973. (That these events were uncoordinated is amply illustrated by the behaviour of Iran which played no part in the embargo yet was a leading hawk on prices.)

Meantime, the questions of participation and the renegotiation of concessions remained unresolved. Notwithstanding the salutary lessons of the CIA-sponsored coup in Iran, there had been sporadic attempts to contest the majors' position from the mid-1950s onwards: in 1956 Egypt nationalized Shell's operations; in 1958 Syria nationalized the Karatchok oilfields while in 1963 the Ba'ath Party took complete control over the oil sector; and in 1967 Algeria began a process of nationalizing US interests, extended to French concerns in 1971. Similarly, Egypt and Iran began joint ventures with the oil companies in 1957, Iran developed the concept of service contracts and Indonesia instituted production sharing. None the less, the increasing nationalist demands for state control of the oil sector faced the pro-Western regimes with a sharp dilemma: refusal to contest the prerogatives of the majors might be taken as a sign of political weakness but nationalization could undermine their alliances with the parent-states of the companies. Yamani's revival of the idea of participation in June 1968 was designed to address this problem, but it was unenthusiastically received by both the majors and the producer-states. Once again, it was the radical regimes acting outside of OPEC, and in the case of Iraq with Soviet assistance, that took the lead. For in the midst of continuing negotiations between the Arab Gulf producers and the majors over participation, on 1 June 1972 Iraq nationalized the IPC, and on the 9th an extraordinary conference of OPEC declared that nationalization was 'a lawful act of sovereignty'. At this point Iran broke ranks, with the Shah announcing that he had concluded a separate deal, widely seen as an attempt to damage the Saudi position on participation in the context of their rivalry in the Gulf. In December 1972 the General Agreement on Participation was signed by Saudi Arabia and Aramco – but it was correctly seen as 'too little, too late'. Indeed, more momentum had already built up behind participation/nationalization when Libya, this time in alliance with the Italian state oil company, ENI, implemented a deal in September 1972.

Despite these advances many OPEC members retained clear links with the Third World, the Non-Aligned Movement and the Group of 77, for only Saudi Arabia, Kuwait, Qatar and the United Arab Emirates received income far in excess of domestic demands. Recognizing the consequent potential for internal division, the Algerian president, Houari Boumedienne, attempted to use OPEC's strength to press for the New International Economic Order. As Terzian explains: 'The danger, as [Boumedienne] saw it, was that OPEC would be brought back into the fold by the industrialized countries, reintegrated into their system by being granted a specific role as crude producers.'[93] This fear was amply confirmed, for until the collapse of the 'North–South dialogue' in the summer of 1977, 'Saudi Arabia systematically strove to avoid any discussion with any bearing, however remote, on the fundamental issues'.[94] This behaviour reflected the triangle of dependence which emerged whereby OPEC was policed by Saudi Arabia while the Kingdom was increasingly aligned with the United States. Meantime, on a shopping trip to Washington for military hardware and nuclear power plants (November 1977), the Shah of Iran made clear his sympathy for price moderation. By these means, together with the conjoint dependence of Iran and Saudi Arabia on the United States, arms sales provided the basis for Riyadh–Tehran leadership over OPEC (the combined production of these members amounted to 48 per cent of OPEC output in 1976–7). And, as noted above, Yamani now tried to link the future movement of the oil price to that of industrial goods in the West. Significantly, this proposal was also adopted by the Trilateral Commission report, *Energy: Managing the Transition*, published in June 1978.

However, steady inflation and the sharp decline of the dollar from July 1977 into 1978 eroded the real value of oil earnings and assets, while the opportunities for productive investment in many OPEC countries increased. In addition, the cartel continued to exhibit economic and political divisions and at the December 1976 meeting of OPEC there was a temporary split as Saudi Arabia tried unsuccessfully to flood the market to keep prices down. Indeed, by 1978 even Saudi Arabia was running a balance of payments deficit on its current account. (In fact, the 1978 price of $12.70 per barrel was only worth some $7 per barrel in $1973; the Saudi strategy had thus moderated the 1973–4 price increases to the extent that the real price of oil had only doubled between October 1973 and 1978). In response to this and the loss of Iranian exports of nearly 4 million barrels per day, due to the strikes in the industry from October 1978, the Saudis increased production: in September Iranian output was some 6 million barrels per day, and by December it had fallen to 2.4 million barrels per day. In this way, at the OPEC conference of December 1978, the Kingdom was able to limit price increases for 1979 to 10 per cent. But the Iranian Revolution of 1979 and the Gulf War beginning in 1980 were to destroy this carefully crafted strategy.

The termination of Iranian oil production during the Revolution resulted in British Petroleum, pleading *force majeure*, stopping third-party sales to other majors and refiners. (BP had received 40 per cent of its total supplies from Iran.) As a consequence, many refiners turned to the nascent spot market to ensure supplies and so avoid losses; some companies also had excess refining capacity to fill. Spot prices increased and the producer-states broke ranks. At the March 1979 OPEC meeting the imposition of 'temporary' surcharges was agreed and the fifty-fifth conference in December took no decision on price, breaking any effective control over pricing: by December 1980 the real value ($1976) of Saudi marker crude was $18.8 per barrel, up from $9.28 in 1977–8. The Saudi Arabian increase of production improved the position of Aramco but did little to ameliorate that of the Japanese companies, BP or Shell who relied on Iran.[95] Moreover, the generalized uncertainty (in July the United States announced it was establishing the Rapid Deployment Force, and in November the US Embassy in Tehran was seized and hostages taken while in Saudi Arabia the Grand Mosque in Mecca was occupied), combined with the attempt to restock for the winter of 1979–80 what had been drawn down in 1978–9, aggravated the volatility of the spot market.

In September 1980 Iraq declared war on Iran and by October the former's oil production had been severely curtailed; another 3 million barrels per day was immediately taken off the market. Elsewhere, however – in Mexico, Egypt and in the British and Norwegian sectors of the North Sea – production was increasing. In addition, demand had been falling, Saudi Arabia increased production and the oil companies had acquired larger inventories. (OPEC's share of production for Western demand fell after 1973 as world demand increased by only 2 per cent per year until 1979. Thereafter, until 1983, oil demand fell by 5 per cent per year and OPEC, being the swing producer, saw its market share fall to 35 per cent – by 1982 the OPEC states were again running a current account deficit.) As a consequence, in marked contrast to the panic spot purchases of 1979, spot prices, after rising for a few months from late September to mid-November, stabilized and thereafter fell. Market sentiment was placated also by a greater confidence in Saudi behaviour as the Kingdom was no doubt assumed to have drawn closer to the United States under the pressure of Iran and the Soviet invasion of Afghanistan in December 1979.

Thus instead of viewing OPEC as a cartel-like assertion of producer-power, it is more accurate to place these developments within the context of the rise of radical nationalist and revolutionary regimes during what Halliday has identified as the 'second wave' of Third World revolutions.[96] In this process, decolonization and formal, independent statehood were accompanied by an upsurge of national and social struggles for development. And so within OPEC, as we have seen, the running was made not by the organization acting in concert but by Algeria, Libya and Iraq with some assistance from Egypt and Syria. And while Iran for a while adopted an intransigent line on prices, essentially to finance the fourth (1968–72) and fifth (1972–8) development plans as well as its massive military buildup, it soon joined Saudi Arabia in policing the organization in the name of moderation. Albeit with the Kingdom playing the dominant role, these states together checked the radical thrust of other members. In this task they were aided by three factors: first, along with the other Gulf producers, Iran and Saudi Arabia controlled the vast bulk of OPEC reserves, thus giving them a freedom of manoeuvre denied to those with lower per capita capacities; second, as the twin pillars of US strategy in the region, both were the privileged recipients of US military hardware, acquired for domestic and regional missions; and third, each state was also tied ot the West to the extent that their domestic state-led capitalist development required extensive economic ties to the core capitalist states for continuing success. Of course, the Iranian Revolution together with the subsequent Iran–Iraq War (1980–8) undercut the basis of this stabilization, as did developing tensions in the US–Saudi Arabian alliance.

The International Energy Agency and US Policy

In addition to the US response to the 'energy crisis' just described – namely, monetary manipulation and an attempt to influence Saudi Arabian and Iranian policy – and in the face of difficulties over domestic reform a third strategy was launched, the establishment of the International Energy Agency (IEA) in 1974.[97] The Nixon administration was concerned for the hold of the US majors over world oil, and recognized the danger of direct arrangements between consuming nations and producer-states which could bypass the US-dominated industry – that is, the generalization of something akin to the traditional French and Italian strategies.[98] And indeed many West European states and Japan quickly dispatched diplomatic trade missions to the Middle East. By July 1974 a Euro-Arab consultative council had been established between the European Community and the Arab League (while US fears were further aroused by the Lome Convention of 1975). Given its general postwar subordination to US policy, the role of Japan was perhaps most surprising: 'Japan . . . endorsed the Arab standpoint in November 1973 and, having once repudiated the American example, continued on this approach by seeking bilateral associations with the producers. Japan promised economic and technical assistance to the oil producers, borrowed petro-dollars, and encouraged Japanese firms to build up exports to OPEC countries'.[99]

But by 1975 the producer-states had established a relatively controlled supply of crude and were thus able to impose competition between the consumer-nations even for 'special arrangements', thereby undermining the point of these putative deals. In addition, the various OECD states had differing interests with respect to OPEC and so the Western response was weakened by endemic conflicts of interest, as well as by any short-term absence of planning or foresight.[100] There were countries such as Japan, Italy and Denmark, which wanted lower oil prices but were none the less heavily dependent on OPEC. Next came those states, including Germany, France and Benelux, which could supply their own solid fuel requirements but relied on imports for oil. Among this key group there was an additional, important division: France favoured a coherent EC response but Germany fell in with the Atlanticist strategy under intense US pressure based on the bilateral security relationship. Then there were the two emergent oil producers, Britain and Norway, both anticipating self-sufficiency in oil, which therefore had no clear interest in large-scale price cuts, at least in the medium term. And finally, there was the United States itself. In 1970 the US balance of trade in fuels was $-1.3 billion, by the early eighties it was $-73 billion; the respective figure for the other components of the balance of trade were $4.4 billion and $51.0 billion.[101] But this dependence did not translate into an unambiguous desire for price cuts because of the role of the US majors, geopolitical considerations in the Middle East, the issue of domestic price controls, the rising trade wars with Western Europe and Japan, the existence of high-cost alternatives in the United States and the ability of the United States to pay for imports in its domestic currency. Let us consider each of these in turn.

The majors were to have a central role in carrying out any emergency allocation programme under the auspices of the IEA, while the US administration gave them the necessary anti-trust clearance. Moreover, the minimum floor-price arrangements, investment incentives, taxation proposals and export provisions of the IEA were designed to protect the investments of the majors and other (largely US) energy corporations against a fall in the price of oil.[102] The US representative to the IEA, Thomas Enders, put the position as follows: 'Each of the consuming countries faces an enormous task in mobilizing the capital required to diminish its dependence on imported energy. . . . But consuming countries can assist each other by making it possible for countries that are capital-rich but poor in energy resources to invest in large-scale projects in other consuming countries'.[103] Such a strategy would not only involve the sharing of the OECD adjustment between the United States, Japan and West Germany but would also give considerable emphasis to the US oil and energy corporations.

The geopolitical considerations, which helped to buttress the argument for a floor-price, were put in equally stark terms by Henry Kissinger:

The only chance to bring oil prices down immediately would be massive political warfare against countries like Saudi Arabia and Iran to make them risk their political stability and maybe their security if they did not cooperate. That is too high a price to pay even for an immediate reduction in oil prices. If you bring about an overthrow of the existing system in Saudi Arabia and a Khadaffi take over, or if you break Iran's image of being capable of resisting outside pressure, you're going to open up political trends that could defeat your economic objectives.[104]

So far, so much rationalization. What Kissinger omitted to say was that the United States had no serious interest in greatly reduced oil prices, even for economic reasons, nor did he point to the positive geopolitical interests that the price rises served. Further, bellicose threats to take the oilfields by force were not well received in the Arab world, to say nothing of the insuperable practical difficulties involved.

Within the US market, federal price controls have a long and detailed history that need not detain us here. Suffice it to say that the effects of such controls were to keep US domestic prices below international levels during the 1970s such that the average cost of oil in the United States was some 40 per cent below world market levels in the first half of 1979.[105] This policy was closely related to the issue of intra-imperialist rivalries and congressional opposition to the removal of controls for reasons of domestic political support. The United States could pay for its imports in its own national currency and it was, therefore, able to escape the balance of payments constraints that its rivals faced. Furthermore, the relative price advantage of US oil gave domestic industry a competitive advantage over Western Europe and Japan. Finally, notwithstanding price controls, higher energy prices made alternative sources increasingly attractive, and in this area – particularly tar sands – the United States possesses large reserves both relatively and absolutely. Thus, although the United States had little difficulty in establishing the IEA to pool OECD resources against OPEC, the Agency has acted not to control the market or even to adopt policies to reduce or stabilize oil prices – witness the disastrous behaviour of the OECD states in the second price increases, artificially driving up spot prices which OPEC obediently followed – but to encourage national policies and develop its own internal arrangements to insure against the effects of any potential disruption to supply.

What of the future? In the early 1980s, Roncaglia argued that the development of the industry could be characterized by any one of three hypothetical scenarios.[106] First there was a 'crisis scenario' in which military, political or ecological crises produced a cut-back in production, in turn producing severe shortages and large price increases. Second, there was the 'Saudi hypothesis' or 'downward stability scenario' wherein a gradual fall in prices occurred over the short term, followed by the 'Yamani formula' of indexing crude prices to the nominal national income of the industralized countries. Third, there was a 'breakdown scenario', resulting either from a collapse of OPEC or a drastic fall in demand or increase in non-OPEC production, in which prices collapsed. We shall see how the 1980s turned out and consider the likely prospects for the 1990s in chapter 6.

Conclusion: Strategies for the 1980s

As we have seen, by the late 1960s developments in the international oil regime, the ending of the long boom and the reshaping of US hegemony resulted in a conjuncture where the oil market moved from consumer- to producer-dominance. At the same time, the rise of radical regimes challenged Western dominance over the oil industry, making real but limited gains until Saudi Arabia and Iran imposed a degree of 'moderation'. In the process, the strategies of the majors, the producer-states and the United States shifted significantly. The consequent restructuring of the international oil industry preceded and postdated the 'OPEC price rises' – gradually the outlines of new strategies became clear.

The majors took advantage of the price increases both to finance and make profitable the acquisition of new reserves in politically safe areas and to re-equilibrate their vertical integration. The higher oil prices of the 1970s and accompanying surges in company profitability financed and made economic the development and production of non-OPEC oil and non-oil energy sources. In turn, this strengthening of the leading oil companies and the broader economic implications of the oil price rises – the expansion of the offshore markets and the growth of unregulated banking capital – reinforced the political bases of US unilateralism. For after a prolonged and deep international recession, the highly unstable patterns of US accumulation during the mid- to late 1980s rested upon the strengthening of bank, oil and rentier incomes in the metropolitan core which in part derived from the oil price rises.

Turning to the actions of the US state, while it is true that the United States lost its former direct control over the international oil regime, its continued predominance within the world's energy markets, and in particular US capital's dominance among global energy corporations and oil-related banks, allowed considerable room for manoeuvre. This position was strengthened by the asymmetrical character of the US energy market: for on the basis of a relatively low level of energy import dependence (though absolute US imports are of great significance for the international energy market) and the ability to pay for imports in its (floating) domestic currency, the United States responded to its loss of control over the oil order with an increasingly *unilateralist* energy strategy. Further, the US concern with regional management in the Middle East, combined with the exploitation of tutelage *vis-à-vis* Western Europe, enabled it to dominate IEA-style strategies. In this manner, US policy largely set the terms through which the IEA member states coped with the consequences of the oil shocks. The priorities adopted were to diversify oil supplies so as to reduce dependence on OPEC, to substitute other fuels for oil – especially nuclear power, gas and coal – to conserve energy in industrial, transport and household sectors, and to continue research into alternative or renewable energy sources. In practice, the first two objectives have been paramount: attention to conservation has been highly variable, while alternatives and renewables were given little priority by the major industrial nations. Thus, if tutelage disciplined the European allies, what imposed a limited unity on the policies of the United States and the other IEA states was their common dependence on the predominantly US international energy and mining corporations, alongside their emphasis on nuclear power and the expansion of internationally traded coal.

In this way, US manoeuvres in the oil sphere paralleled its wider response to the alteration of hegemony during the 1970s; however, it is not at all clear that the policies have been unambiguously successful. For although the United States played with decisive advantages in bilateral relations in the energy (and other) spheres, unilateral actions with respect to the Middle East and Western Europe have arguably been counter productive, in so far as the attempt to construct a new phase of hegemony under the banner of a Cold War offensive against an 'expansionist' Soviet Union has had ambiguous consequences for the long-run position of the United States in economic, political and energy (if not military) terms. However, the opposition of the United States did make the search for a unified West European energy policy much more difficult. And Western Europe was further antagonized by the US refusal to pursue a policy of deep price cuts in oil, a stance dictated by the central role of the majors in the international energy market, geopolitical considerations of stability in the Middle East, rising trade wars in the OECD bloc and the existence of high-cost oil and alternatives in the United States.

In this context, the 'loss' of Iran was particularly significant in that it demonstrated the precarious nature of oil-based and foreign transfer financed 'modernization', while simultaneously considerably weakening the US position in the region. Using the threat of incipient Soviet expansionism, the United States accordingly tried to build a regional anti-Soviet coalition, strengthened strategic ties with Israel and Saudi

Arabia, promoted open trade and investment and increased the direct US military presence. (The Reagan corollary to the Carter Doctrine should be noted here.) Within OPEC itself the central role has continued to be that played by Saudi Arabia, whose strategy consists in aiming for high unit royalties as opposed to high volumes of production, using their market control to avoid erratic supply and price fluctuations, and maintaining their role as the leader in OPEC. To this extent the Saudi strategy has not been inconsistent with that of the United States and the other IEA states, and the United States has sought to build on this and its military and economic connections to Riyadh to maintain a high degree of control over the international oil market. These themes will be returned to in chapter 6.

5

Allies and Rivals: Petropolitics in Europe, Japan and the Soviet Union

Introduction

As we have seen, throughout the history of the global oil industry, European states, Japan and the Soviet Union have all played greater or lesser roles from time to time. Thus far, however, this study has focused mainly on the United States, the Middle East and the leading international oil companies. This chapter seeks to redress the balance a little by situating the responses of these states during the 1970s and eighties in a broader context. But rather than provide a detailed account of oil and energy policy in each case, the following treatment will look at the postwar period and the relations between these blocs and the US-dominated international industry. Throughout, attention will be drawn to the way in which the organization of the oil industry has been related to the broader skein of US hegemony, with its multiple connections to the Cold War and intercapitalist rivalries. Specifically, this chapter seeks both to explain the divergence of Western positions around the events of 1973–4, as well as the continuing problems energy debates have posed to the Atlantic alliance, and to explain why the considerable Soviet petroleum reserves have not served as a stable basis from which to project international power.

Accordingly, I consider the position of (1) Western Europe, (2) Japan, and (3) the Soviet Union. Once the position of these regions in the political economy of international oil has been sketched, (4) I conclude the discussion by attempting to draw together the ways in which each of the regions is imbricated in and circumscribed by the world oil order constructed after the Second World War. In terms of the conjunctural model developed thus far, this chapter argues: first, that although the power derived by the United States from its direct control over European and Japanese oil has been weakened, the continued structural dominance of the United States in the international system has largely offset this; and second, that because of its structurally weak position in the world economy, the Soviet Union has been unable to project significant global power through its petroleum policies. Thus, what follows also serves to provide the necessary backdrop for chapter 6, which considers the responses of the United States and the majors to the upheavals of the 1970s and the resultant re-ordering of world oil.

Western Europe

The European Community, Energy and Oil

Although the establishment of the European Coal and Steel Community (ECSC) in 1950 had been an important forerunner to the broader Community, the formation of the European Community in 1958 did not lead to the development of a unified energy policy among its members.[1] For a start, responsibility for different fuels was divided between separate agencies: the Paris Treaty allotted coal to the ECSC; the Treaty of Rome gave oil, natural gas, hydropower and electricity to the Commission; and nuclear power came under the auspices of Euratom. Moreover, none of these arrangements included provisions for the development of a common energy policy. A yet more basic obstacle lay in the fact that most of the components of Western Europe's energy industries were state-owned or -regulated, nationally based companies. For this reason alone, any attempt at rationalization would have engendered complex disagreements relating to regional unemployment, and national claims of sovereignty. Indeed, given the priority accorded to national reconstruction during the 1950s, and the subsequent moves towards corporatist planning in the 1960s, many states regarded energy as a key sector for *dirigiste* coordination. Equally, given that the most powerful members of the Community – West Germany, France and Italy (and later the United Kingdom), as well as Belgium, Denmark and the Netherlands – were members of NATO, and given also that while the Community aspired to a common political definition, it played little role in the field of foreign policy; such emergency planning as did take place for the security of oil supplies in times of crisis and war occurred in NATO councils, rather than in the EC.

Still, perhaps the most important reason for this neglect of a common energy policy was that no such policy appeared to be necessary. In the 1950s and sixties the central concerns in West European energy matters were the unemployment and regional implications of the rundown of the coal sector and the security of rapidly expanding crude oil imports from the Middle East. After the cautious outlook of both the Armand (1955) and Hartley (1956) reports, the Robinson Report of 1959 foresaw an energy surplus and took an altogether more sanguine view of oil imports. And as the oil companies demonstrated that their world-wide distribution systems could cope with the supply disruptions of the Suez Crisis (1956) and the 1967 Arab–Israeli War, fears over security of supply were soon dissipated. The results were dramatic: in 1950 coal accounted for 75 per cent of the European Community's total primary energy requirements and oil some 10 per cent, by 1966 the corresponding figures were 38 and 45 per cent and in 1971 20 and 60 per cent, respectively. The Community's energy import dependence increased in parallel: in 1950 imports amounted to 13 per cent of requirements, by 1960 this rose to 30 per cent and in 1970 it stood at 63 per cent.[2] Finally, for the longer term, the prospects for nuclear power seemed to be assured, such that imported oil would provide a temporary bridge between a coal-based economy and the anticipated large-scale shift to nuclear-generated electricity. Under these circumstances, the main community policies related to the regularization of coal subsidies, the harmonization of taxes and the promotion of competition.

Nevertheless, the displacement of coal by oil cannot be explained simply in terms of the movement of relative prices in an increasingly open market. For, as a direct consequence of the West's economic and strategic dominance in the Middle East together with the connected multinational control of the region's oil, Europe's imports were priced well below the opportunity cost of available alternatives. Furthermore, the terms of trade between coal and oil were regulated both by the oligopoly control of the majors and the fiscal policies of the consuming nations. In combination with their unilateral ability to determine prices, the oil companies profited from the presence of refining capacity in Western Europe, and on this basis set petrol prices considerably higher than those for fuel oil. The latter, which in many cases was an alternative to coal, 'was being dumped, while gasoline was being monopolized':[3] while costs of production differed only marginally, petrol prices ranged from two to seven times those of fuel oil. Government policy

TABLE 5.1   Shares of coal and petroleum (oil and gas) in energy demand (%)

	1955		1980	
	Coal	Petroleum	Coal	Petroleum
Britain	85	14	37	58
France	61	28	18	65
West Germany	88	9	30	64
Italy	25	44	8	85

Source: Adapted from Nigel Lucas (with the assistance of Dimitri Papaconstantinou), *Western European Energy Policies*, Clarendon Press, Oxford 1985, table 4.1, p. 141.

TABLE 5.2   Total energy demands by fuel in 1973 (%)

	Oil	Natural gas	Coal	Others
Britain	52.1	13.2	33.6	1.2
France	72.5	8.1	16.1	3.2
West Germany	58.6	10.1	30.1	1.3
Italy	78.6	10.0	8.6	3.2

Source: Adapted from Romano Prodi and Alberto Clo, 'Europe', in Raymond Vernon (ed.), *Daedalus: The Oil Crisis*, 104, 1975, table 4, p. 95.

routinely compounded the situation in two ways: first, restrictions on investment in the coal industry left it in a weak competitive position during times of shortage; and second, fiscal policy supported the strategy of the majors, to the extent that duties were higher on petrol than on fuel oil.

As a result, West European dependence on imported oil rose rapidly and, despite the growth of independents and national oil companies in the 1950s and sixties, the Seven Sisters controlled some 65 per cent of its supply in 1973. By country of origin, Community oil imports in 1972 were as follows: Saudi Arabia 23 per cent, Kuwait 14 per cent, Libya 14 per cent, Iran 11 per cent and Nigeria 9 per cent.[4] For the overall picture, an indication of the changing energy balances within the Comunity and those immediately prior to the 1973–4 price increases can be seen from tables 5.1–5.3.

TABLE 5.3 Domestic production as proportion of domestic consumption by fuel in 1972 (%)

	Oil	Natural gas	Coal	Total
Britain	2	97	98	51
France	1	54	69	23
West Germany	7	64	115	50
Italy	1	93	5	15

Source: Adapted from Romano Prodi and Alberto Clo, 'Europe', in Raymond Vernon (ed.), Daedalus: The Oil Crisis, 104, 1975, table 5, p. 95.

National Oil Companies in Europe

The domination of the majors in West European oil supply and processing had not gone unchallenged, however. Immediately prior to the First World War, strategic considerations relating to fuel for the Royal Navy prompted the British state to render assistance to the Anglo-Persian Oil Company, thus consolidating the position of the latter in Iran. And during the interwar period a combination of strategic pressures and economic incentives resulted in other European states encouraging the formation of national champions in the oil sector. Thus, as Samuels has noted, 'all state-owned oil companies were formed as commerical ventures justified by strategic national goals in the face of initial market domination by a small number of privately held, usually foreign multi-nationals'.[5] None the less, although those European powers active in the Middle East joined (rather than challenged) the US-dominated system based on the majors – Britain established a major (BP) and France's Compagnie Française de Petrole (CFP) can be regarded as an honorary member of the club – closed production and trading circuits dominated the world oil industry from 1928 to 1973. For in circumstances where European power in the Middle East had been displaced by that of the United States, the growing strategic nature of oil implied that US-controlled oil, sourced by the majors from the Middle East (and later North Africa), would underpin the economic boom resulting from the postwar phase of reconstruction. Moreover, if the administration of the Marshall Plan facilitated this shift from a coal- to an oil-based economy (Walter J. Levy came from Mobil to head the oil division of the Economic Cooperation Administration), the economies of Western Europe still derived considerable benefits from the cheap and (in real terms) falling price of oil.[6]

The most successful instance of the formation of a national champion, of course, was the early assistance given by the British state to the formation of the Anglo-Persian Oil Company (later BP). In this case, a coincidence of interests between the Admiralty and private oil capital, in opposition to both Shell (considered unreliable because of strong German influence in Holland) and the domestic coal industry, resulted in the formation of a powerful, vertically integrated major during the First World War. Shortly after the war, the leading national oil company in France, CFP, emerged from a state-orchestrated merger of private capitals in circumstances of acute competitive pressure from the majors. This was followed by an aggressive move into refining which, in combination with the oil import law of 1928, protected the French market until the late 1970s. After the First World War, France acquired German interests in the reformed Turkish Petroleum Company and maintained colonial spoils in Algeria. Thus the CFP was well placed in Iran, Iraq and Algeria, and has often been regarded as the eighth major. But the postwar spread of state control in the Middle East, together with the Iranian Revolution and the Gulf War, cut supplies back sharply and new sources were secured from Mexico, Venezuela and Norway.

A marked difference is indicated by the ambitious Italian attempt to break the power of (rather than join) the majors. Excluded from significant imperialist possessions, the Fascist economy created a national company, Azienda Generale Italiana Petroli (AGIP), in 1926, 'formally established as a parastatal body to assist domestic industry and commerce, to control the Italian oil market, and to end market domination by foreign trusts'.[7] At this time, Standard Oil and Shell controlled three-quarters of Italian oil production and an even greater share of refined products. By 1933 Standard Oil still controlled nearly 75 per cent of the Italian market (much of the balance came from the Soviet Union). In 1953, however, Enrico Mattei ensured that AGIP was restructured into a new company, Ente Nazionale Idrocaburi (ENI), which was given a monopoly over the Po valley gas reserves; and ENI was soon producing 90 per cent of the country's natural gas as well as marketing 25 per cent of its oil products. In addition to Italy's own rapid shift to a heavy reliance on oil as both the dominant primary fuel and the leading input for the conversion industries, its geographical location between the Middle East and the markets of Southern Europe and Germany 'made it a natural choice as a refining centre'.[8] Outside the West European markets, ENI signed deals with both the Soviet Union (1959 and 1961) and with radical, independent states in the Middle East (such as Iraq). But its ambitions were curtailed when, with Mattei's death in 1962, the secret financial contributions made by the majors to parties and politicians increased, thereby limiting ENI's freedom of manoeuvre. In addition, US pressure through NATO placed a ceiling on Soviet imports.

The West German market, by contrast, is probably the most open in Western Europe. As a result of defeat in First World War, and the consequent loss of its colonies, German industry turned to the synthetic production of oil and struck deals with Romania and the Soviet Union. The destruction of domestic capacity during the Second World War, together with the postwar reorganization of the industry under Allied occupation, resulted in an oil market unimpeded by restrictions on entry, prices, investment or foreign ownership. By 1950, 83 per cent of West German crude imports originated in the Middle East with most of the balance coming from Venezuela. During the 1960s, pressed by competition from ENI, the majors constructed pipelines from Marseilles and Trieste into the Federal Republic allowing a rapid increase in imports from North Africa: by 1970, 59 per cent of total imports came from Africa, with 41 per cent from Libya alone. Another shift came with the expansion of North Sea exports in the 1980s, and by 1984 this region accounted for 30.5 per cent of crude imports.[9] Concerned by its dependence on the majors, in 1969 the government formed a state oil company, Deminex, which was reorganized in the mid-seventies under the control of Veba: but by the 1990s, Veba will control a mere 10 per cent of total consumption. Meanwhile, taking advantage of the economic *détente* of the 1970s, West Germany also played a leading role in securing imports of Soviet gas through complex barter deals. On the other hand, West Germany has argued for price and production transparency within the International Energy Agency (IEA) and has strongly supported proposals and plans for sharing schemes during any future emergency. What these otherwise diverse policies have in common is that the export strength of the German economy has enabled it to adapt to the vicissitudes of its external exposure. Lucas has ably characterized this underlying continuity in the German approach as follows:

West Germany understands that energy is too important a matter to be left to energy policy. It is not a matter of substituting new political relationships for ancient concessions in oil rich states and of substituting new energy forms for oil; it is also a question of adjusting industrial behaviour to a quite different distribution of economic power.[10]

The European Community and the Oil Crisis

Overall, then, the events of 1973–4 found the members of the European Community with widely differing strategies towards oil and without an agreed energy policy, to say nothing of forms of emergency provision. Thus, in circumstances of a high and growing dependence on Middle East oil, the Community's response was uncoordinated and divided. However, contrary to common belief, neither Western Europe as a whole nor the members of the Community suffered an overall shortage during the Arab embargo:

> So far as Europe was concerned, the oil crisis did not bring on an overall shortage in either oil or energy. The difficulties of the crisis were mainly the result of disturbances in the internal distribution of oil products, which were triggered by the resistance of governments to the price increases imposed by the oil companies, and which led to delays in delivery, discrimination against independent companies, and the speculative hoarding of stocks.[11]

Nevertheless, the price rises and the supply disruptions dramatically exposed the consequences deriving from the absence of a Community strategy. In response, in 1974 the Commission presented the Council of Ministers with a long-term strategy paper, *Problems and resources of energy policy, 1975–1985*, in which it advanced the case for a unified internal market for energy alongside policies to cope with security of supply. The latter was to be pursued by increasing electricity's share of final consumption from 25 to 35 per cent, by a general increase in the use of nuclear power, coal and natural gas, and therefore by reducing the share of oil from some 60 to nearer 40 per cent of total energy requirements. But although the Commission advocated a unified energy market, it also urged the coordination of a 'flexible system of concertation', involving the relevant energy corporations and the Committee of Energy. None of these proposals was backed by a concrete programme of action.

In fact, any such strategy would have soon run into the divergent stances and interests of the member states. Most countries, especially those with an existing presence in the international oil industry such as Britain (BP and Shell), France (CFP and Elf-Erap) and Italy (ENI), attempted to secure stable supplies through bilateral deals. Meantime, attitudes differed towards the US-sponsored IEA: France argued for an independent EC dialogue with the producer-states; West Germany, Denmark and the Benelux countries encouraged a common Community policy but did not support the anti-US stance of the French; and Britain, Italy and Ireland favoured no action at all. All except France, however, were reluctant to compromise Atlantic unity by obstructing US proposals. Therefore, at the Washington Conference of February 1974 all the members of the Community – again except France – lined up with the United States to combat the power of OPEC. Equally, the influence of majors (BP and Shell) in Britain and the Netherlands, combined with the growing importance of North Sea reserves to these states and Norway, meant that there was a powerful block which would refuse the definition of a common EC policy, let alone one for Western Europe as a whole. Thus, the combination of US pressure and European disunity resulted in the member-states of the EC participating in the US-sponsored IEA, rather than acting coherently as a bloc.

Notwithstanding this lack of an agreed policy within the EC, a number of summits resulted in target plans for the future shares of different fuels.[12] Between 1973 and 1990 coal's share was projected to increase slightly (23–24.7 per cent), oil decline rapidly (59–42.4 per cent), natural gas increase sharply (12–18.1 per cent), nuclear power expand vigorously (1–12.5 per cent) and others decline marginally (3–2.3 per cent). And although the ten Community members were already some 44 per cent energy dependent in the mid-1980s, the subsequent expansion of the Community added two more energy-deficit countries, Spain and Portugal. In addition, within the enlarged Community, only the UK (103.5 million tons), Germany (85), France (17), Spain (15) and Belgium (6) are coal producers.[13] This overall EC deficit, in conjunction with the rundown of European deep-mined coal, does indeed indicate that the Community is planning both to increase the role of electricity in final energy consumption and to rely upon imported coal and a considerable expansion of nuclear power for this new capacity. But after the public reappraisal of nuclear power in the wake of the Chernobyl disaster, its prospects in Europe outside of France seemed bleak, a predicament compounded by the gradual removal of the massive state subsidies formerly paid to this sector as privatization and deregulation spread.

Thus, with the international trade in steam coal expected to expand rapidly, and with the EC presently taking over one-third of sea-traded coal (up from 41 million tons in 1975 to 95–8 in 1988), indigenous European production may well continue to fall – already between 1975 and 1985 some 50 million tons of capacity was lost. Moreover, the trend towards the closure of deep-mined capacity and increased coal imports might be augmented by the continuing completion of the 'internal market' (1992). For although the Commission has moved

slowly in this sphere, it has begun to consider freeing EC energy
markets by looking at the subsidies provided to West German deep-
mined coal through the *Jahrhundertvertrag* and the *Kohlepfennig*.[14]
More generally, the Commission has argued that the benefits of a
completed internal market in energy can only be realized if the Community adopts a common fuel policy, moves towards a system of
common carriers and relaunches nuclear power early in the next century.

Britain and Norway Go It Alone

As suggested above, both Britain and Norway jealously guarded a
degree of national control over their North Sea production and reserves,
though the form that this took differed significantly. The rapid exploitation of petroleum in the British case was determined above all by a
desire to escape the balance of payments constraint on domestic
demand management.[15] This resulted in a high level of dependence on
inward, multinational (predominantly US) capital to exploit the reserves
and an accompanying loss of control over the industry by the state.
Ambitious plans to increase public control – including the proposed
National Hydrocarbons Corporation in the 1960s and the Labour policies of 1974–9 – were thwarted by the overriding concern of both
the Conservative and the Labour parties to proceed rapidly, thereby
promoting the growing multinational penetration of the British economy. As a consequence of this disintegration of a social democratic
alternative, petroleum rents were harnessed to the monetarist and
neoliberal strategies of the Thatcher administrations, providing finance
for the Treasury and funding overseas investment. To this end, production of oil and gas were maximized and in the case of oil the
highest market price sought. The abolition of the British National Oil
Corporation (BNOC) and the privatization of Britoil, alongside the
privatization and deregulation of British Gas, further opened UK markets to transnational capital.

The case of Norway has been altogether different.[16] Because of the
potential size of Norway's petroleum economy as compared with its
other sectors, the management of this resource required long-term
planning if rapid and destabilizing structural adjustments to the main-
land economy were to be avoided. Moreover, there was no pressing
balance of payments constraint. Still, the capital and technical requirements of exploration and development were beyond those of the domestic private sector, so either the international oil companies or the

TABLE 5.4 Crude oil production and exports of UK and Norway compared to other producers in the noncommunist world, 1984

	Production		Exports	
	mb/d	%	mb/d	%
United Kingdom	2.58	4.5	1.55	7.3
Norway	0.71	1.2	0.62	2.9

Source: Adapted from R. Mabro, R. Bacon, M. Chadwick, M. Hallivrell and D. Long, *The Market for North Sea Crude Oil*, Oxford University Press, Oxford 1986, tables 1.3 and 1.4, pp. 7–8.

TABLE 5.5 North Sea and OPEC shares of crude imports, selected countries, 1984 (%)

	US	Japan	Germany	France	Italy
North Sea	11.5	—	30.5	20.2	2.0
OPEC	42.0	77.7	55.6	58.0	68.4

Source: R. Mabro et al., *The Market for North Sea Crude Oil*, Oxford University Press, Oxford 1986, table 4.10, p. 68.

state would have to mobilize the necessary resources. Unlike in Britain, however, there was no large transnational sector of the economy which the state sought to protect, and in particular there was no indigenous oil company of any size; the energy-intensive nature of much Norwegian industry, combined with the planned nature of postwar reconstruction, meant that a coherent energy policy already existed; and state intervention in and guidance of the economy was well developed through credit and labour market controls. In consequence, Norway managed its petroleum production decisions and the related rents such that respectable levels of growth, low unemployment and moderate inflation were all maintained during the recessions of the 1970s and eighties. (Of course, the unanticipated price rises of the seventies produced an economy that became overly dependent on petroleum rents, and the precipitous fall in prices of the mid-eighties presaged the beginning of a long period of adjustment.)

Together, the total oil production and exports from the British and Norwegian sectors of the North Sea represent a small but significant portion of the world market, as can be seen in tables 5.4 and 5.5.

The British production strategy has been enthusiastically received by its IEA partners, but the Norwegian attempt to link production to the social and economic demands of its small population (around 4 million) has been the cause of some concern. For while Norway proclaimed a desire to maintain relations with both the West and OPEC, in practice this involved a passive oil diplomacy between 1973 and 1978. After the second oil shock, however, Norway's partners in the IEA urged greater production in order to reduce dependence on OPEC oil. (Such sentiments were raised in Oslo by the Secretary General of the IEA, Ulf Lantzke, in the Autumn of 1980.) In contrast to this, during 1979–80 both Prime Minister Nordli and the Minister of Foreign Affairs repeatedly spoke of Norway's role in the promotion of a new global energy order. Indeed, the United Nations World Commission on the Environment was chaired by Gro Harlem Brundtland (Minister of Environment, 1974–9). None the less, by 1980–1 this moment was past. Meantime, as the IEA developed, so did Norwegian opposition to joining it. Foreign Minister Frydenlund met with Kissinger and opted for associate status in which Norway refused to participate in emergency planning for oil-sharing during a future crisis. Although the Labour government had wanted full membership, domestic opinion ruled this out. (In the early 1970s the Labour Party had received a sharp setback from the 'No' vote in the referendum on Norway's membership of the European Community.) Rather, Norway made plain its preference for bilateral deals which would not compromise its national control over petroleum production to the extent that membership of a multilateral forum might. In particular, in the mid-seventies the government tried to use its energy surplus to develop industrial co-operation with Sweden and West Germany.

Throughout, Norway has placed a priority on avoiding conflict with its allies, yet in the 1970s and early eightes a set of enforceable national priorities, bolstered by cross-party consensus, enabled it to sidestep US and EC pressure for greatly increased production levels – especially of natural gas. By the late eighties, however, pressure began to mount again to expand gas sales. On the domestic side, the fall in the real price of oil cut Norway's export earnings (the 1985–9 Labour government tied Norway's prices to those of OPEC), thus strengthening the economic case for increased gas exports. And among Norway's European partners, growing reservations – both economic and ecological – about the long-term future of nuclear power argued for an expansion of gas-fired electricity generation, as in Sweden for example. In fact, Norway possesses about one-half of Western Europe's natural gas reserves

(accounting for some 60 per cent of the country's total energy reserves
and 30 per cent of its petroleum sales by the end of the 1980s).

Conflicts Across the Atlantic

The Urengoi pipeline was 'an economic lifeline from the West which
offers perhaps the final opportunity for the Soviet leadership to bail out
a bankrupt and failing system'.
        Lawrence Brady, US Assistant Secretary of Commerce for Trade
                                              Administration 1982

Europe 'can't gang up on the United States anymore'.
                                        Richard Nixon, March 1974

We have seen in chapter 4 that Atlantic tensions rose during the
1970s as conflict increased over trade and monetary policies. Indeed,
this lay at the root of French opposition to the US proposals for the
IEA. Continued Gaullist discomfort with the 'inordinate privilege' of
the dollar, combined with France's absence from NATO planning
councils, predisposed it towards an independent stance. Added to this,
the French were exempted from the Arab embargo by virtue of their
position in the Middle East and a large proportion of France's trade
was with the Middle East and North Africa. While these characteristics
shaped the particularity of the French response, general European
sentiment was unsympathetic to the stance taken by the United States.
Calleo ably characterized the position as follows:

In European eyes, American profligacy, having first created conditions
in the oil market that made possible the quadrupling of prices, has
condemned Europeans to a choice between overvalued currencies or
inflation. Because oil is factored in dollars, and Saudi Arabia is an
American military protectorate, nothing restrains the United States from
importing as much oil as it pleases. The resulting 'petro-dollars' either
return to the United States, or make their way to Europe, where they
pose the old dilemma.[17]

Furthermore, successive US administrations were deeply concerned
by the potential West European opening both to the Eastern bloc made
possible by *détente* and to the Middle East in the vacuum that followed
the collapse of the Nixon Doctrine in Iran.[18] The European allies would
now have to be roped into a renewed imperialist offensive to counter
the expansion of East–West trade and the assertion of Third World

power. On the one hand, as superpower relations deteriorated, the United States sought to impose greater controls on Western Europe's economic *détente*. And on the other, with the exhaustion of the oil-profit recycling strategy of the seventies, any potential of Western Europe to pursue an independent *détente* or bilateral and preferential deals with the Third World (especially OPEC) had to be averted. This strategy was to have limited results.

At the start of the Cold War, trade with the Eastern bloc was regulated through the Mutual Defense Assistance Act of 1951 which linked US aid to its allies with compliance on trade controls. But as the benefits of trade with the East grew relative to US aid, and as the reconstruction of Western Europe neared completion, 'the USA's leadership of Western policies on East–West trade began a long process of erosion'; similar considerations applied in the case of Japan as it began to 'cooperate in the exploitation of Siberian raw materials'.[19] Steadily, the share of Western Europe's trade exchanged with the Eastern bloc rose, reaching some 2–5 per cent by 1980, and accounting for some 70 per cent of total East–West trade. The commodity composition of US and European trade with the East also differed markedly: whereas 80 per cent of US exports to Comecon were agricultural products and raw materials, a similar share of Japanese and West German exports were accounted for by manufactured goods, machinery and transport equipment. West Germany's imports from the East, by contrast, included 45 per cent energy (76 per cent in the case of the Soviet Union). By the mid-seventies, the failure of US ambitions for *détente* meant that Kissinger's 'web of constructive relationships' or linkage came under increasing domestic pressure. Thus when Congress passed the Jackson–Vanik amendment in 1974, followed by the Soviet renunciation of the 1972 US–USSR trade agreement in 1975, US pressure on its allies to limit trade and credits to the East mounted.

In the case of natural gas, since the liquefied product could now be transported relatively cheaply and safely, imports from Norway, Algeria, Nigeria and even Iran, together with those from the Soviet Union, became feasible as the integration and deregulation of the European gas industries moved ahead. In fact, after the EC effectively banned new gas-fired power stations in 1975, gas lost one-third of its power generation market in Europe from 1980 onwards. But throughout the eighties world gas reserves increased by about 50 per cent and the Soviet Union (which alone had 43 per cent of total world reserves) was keen to expand its gas sales to Western Europe. This proved to be an area of sustained conflict between the Reagan administration and its European allies. On the pretext of punishing the Soviet Union for the declaration of martial law in Poland, the United States tried to prevent the sale of Western equipment required in the construction of the Urengoi–Western Europe gas pipeline.[20] Although this raised questions about both the international legality of US actions (the Europeans argued that it constituted an attempt to exert extraterritorial jurisdiction) and the consistency of its motives (US grain sales continued), the real conflict lay in divergent attitudes towards East–West trade, as the Urengoi pipeline episode crystallized growing differences within the NATO alliance.

In the 1980s, East European energy exports to the West were set to decline, with the exceptions of Polish coal, Romanian oil and trade between East and West Germany. And in the Soviet case gas was to replace oil; it was reckoned that gas from the Soviet Union might supply 25–35 per cent of West German, Italian and French natural gas requirements by the late eighties (or 5–6 per cent of total energy needs).[21] Not only would this allow diversification away from Middle East oil but it would also offset the simultaneous fall in Soviet oil exports. Nevertheless, in the autumn of 1981, US officials offered the European allies an alternative package which included an 'acceleration of gas supplies from the North Sea countries (especially Norway), Algerian and Nigerian gas supplies (which had been explicitly renounced by the US [on security grounds]), greater imports of US coal and synthetics, [and an] acceleration of nuclear power programmes'.[22] This generous offer was declined. The position of the United States was further weakened by divided counsels within the adminstration: on the one hand, the State and Commerce departments questioned the wisdom of antagonizing the allies in such a direct manner; and on the other, the National Security Council and Defense Department sought to deny hard currency to the Soviet Union, thereby prosecuting economic warfare.[23] Eventually, in November 1982, an agreement was found which lifted US sanctions in return for greater restraint on East–West trade in general and closer monitoring by the Coordinating Committee for East–West Trade. In this instance, the attempt to undercut West European trade with the East had failed. Indeed, in 1989 *The Economist* reported that the Soviet agency Soyuzgazexport was planning to increase its West European gas sales by 40 per cent by 1995. If this goes ahead, Soviet and Norwegian supplies could feed into the expansion of small combined-cycle gas-fired power stations which, one study has estimated, might account for nearly 20 per cent of Western Europe's electricity in 2020 as compared with 3 per cent in 1989.[24]

In the case of oil, by contrast, European initiatives to capitalize on its predominant position in Middle East trade proved unsuccessful. As we have seen, Kissinger's 'Year of Europe' did not turn out as he had intended. In relation to the Middle East, Atlantic relations were greatly strained by the US nuclear alert during the Arab–Israeli War of October 1973, but the ensuing conflict centred on oil. Pressed by the Arabs, in November 1973 the Community issued a joint declaration, calling for a cease-fire and an eventual Israeli withdrawal to its pre-1967 borders. Moreover, the conjunction of US nuclear diplomacy during the war and the differential dependence of Europe and the United States on Middle East oil encouraged a short-lived impetus for an independent Western European security posture. Once again it was a French initiat- ive, with Paris proposing policy coordination in the West European Union (WEU) permanent weapons committee. This was overruled by West German insistence that any discussion take place in the Eurogroup of NATO, in which the French did not participate. Indeed, in general Brandt was anxious to pacify Washington and conciliatory notes were exchanged with Nixon.

At the founding talks of the IEA (February 1974), Nixon linked the question of US troops in Western Europe to better political and econ- omic co-operation across the Atlantic, a point reiterated by Kissinger. Indeed, as Pryce has noted: 'The Washington Energy Conference was a turning point in relations among the allied countries. By uniting against France and with the Americans when the crux finally came, the other Europeans expressed disapproval of transatlantic confronta- tion.'[25] Specifically, West Germany proposed and Kissinger and Nixon agreed that after EC political directors reached a common position, but before the Community's foreign ministers took decisions, Washington would be consulted. In this way, 'only after the Europeans established a consultation procedure and relations with Washington improved could the Euro-Arab Dialogue proceed'.[26] Thus, the attempt by the EC states to secure oil supplies through a Euro-Arab dialogue foundered under the weight of what Lieber has accurately described as 'fierce American opposition'.[27] Equally, there were unresolved problems in the Euro- Arab relation itself: problems of Palestinian representation proved intractable and by ruling out direct discussion of oil pricing and pro- duction levels the Arab League made clear its refusal to compromise the position of OPEC.

None the less, the Middle East's foreign trade is overwhelmingly with Western Europe: between 1977 and 1983 the US dependence on the Middle East for imports was roughly stable, while Japan's depen- dence slightly decreased and that of Western Europe marginally rose; meanwhile, the US share of exports to the Middle East increased from one-fifth to one-quarter of the total, Japan's increased slightly to one-tenth and Western Europe's portion fell a little to two-thirds. In the mid- to late 1980s, the US economic position weakened as oil imports rose and exports (apart from agricultural produce) increased more slowly than those of Japan and the European Community. Thus while the shares of total exports destined for OPEC do not vary greatly between the United States, Japan, and the European Community, the fact that their shares of world manufacturing trade in 1987 were 10.5, 13.0 and 43.1 per cent, respectively, meant that the bulk of OPEC imports came from Western Europe.[28] Moreover, the conjunction of intensified competition in the international arms market, the US embargo on trade with Iran and congressional obstacles in the way of sales to Arab states opened a space for British and French expansion, making Britain the second largest arms exporter in world markets by 1986. But for all this, neither the Community collectively nor any member state alone rivals the overall influence of the United States in the region.[29]

In addition to the failed attempt to block the gas deal and the more successful strategy in the Middle East, a final terrain through which US influence was exercised was NATO. These were closely related issues, for the predominance of European trading interests in the Middle East implied that US influence could be best exercised on the military plane. Also, in order to gain congressional funding for the Rapid Deployment Force, the administration was obliged to raise again the 'out of area' question within NATO. This became a key axis of division within the alliance, with the eventual 'European acquiescence . . . [being] more the product of U.S. pressure than of growing European appreciation of the need for a new approach toward the Gulf'.[30] Nevertheless, the extent of US success was registered by the fact that in late 1987 French, British, Italian, Belgian and Dutch warships provided one-third of the seventy-five NATO warships in the Gulf, while West Germany substituted for Belgian and US forces in the English Channel and the Mediterranean.[31]

For while Western Europe, and the EC in particular, might prefer to deal with the Gulf primarily on the economic plane, thus enhancing its position vis-à-vis the United States, the Western powers shared the higher objective of averting Soviet influence in the region, thereby providing the basis on which US leadership was secured. Thus, given the public and long-standing opposition of Western Europe to US policy in the Middle East in general and the Gulf in particular, the coordination of this activity through the WEU seemed to signal that the WEU would link the EC to NATO as a second pillar of the alliance, rather than provide the forum for an independent Western European bloc. On this occasion at least, interbloc rivalries continued to remain subordinate to the overarching East–West competition for influence in the Middle East.

Japan and the US Oil Order: Two Hungry Giants

On the morrow of Pearl Harbor, the Secretary of the American Asiatic Association wrote privately to his friend Stanley Hornbeck in the State Department: 'It will be a long, hard war, but after it is over Uncle Sam will do the talking in this world.'[32]

If you want to understand the future, you must . . . understand the Pacific region.

(Secretary of State George Shultz, 1983)

Postwar Development

Although there has been an oil industry in Japan since the Meiji Restoration, it has only played a significant role since the Pacific War. During the First World War, Japan eradicated German influence in China and the Pacific region, and European exports to Asia dried up, making Japan and the United States the dominant traders. During the 1920s, Shidehara diplomacy attempted to pursue a policy of co-operation with the Western powers. But as the international situation deteriorated and as the world economy fractured into regional trading blocs, Japan's quest for economic self-sufficiency took a new turn. For in the 1930s the early industrialization based on textiles was beginning to be superseded by a bloc of steel, chemicals, armaments and ultimately automobiles, requiring growing imports of raw materials and expanding export markets. And so, after turning Manchuria into a puppet regime in 1931 and prosecuting an undeclared war in China from 1937, the aim of creating a Greater East Asia Co-Prosperity Sphere was uppermost. Domestically, this line was secured through the ascendancy of the Toheiha faction (or Control Group), both within the army and over politicians, together with its growing links to the *Zaibatsu*. In March 1941 a Neutrality Pact was signed with the Soviet Union, after which Japan invaded Indo-China in order to cut Chinese supply lines. This in turn provoked an economic blockade by the United States, Britain and the Netherlands. Thus, lacking domestic oil supplies, continued military viability demanded a push towards the East Indies. In this context, given that the United States was unwilling to allow Japan to dominate Southeast Asia, conflict was more or less inevitable.[33]

European interests in Southeast Asia were fundamentally and irreparably damaged by the war. Not only was the vast bulk of Allied fighting in the region accomplished by US forces, thereby eclipsing Britain from the occupying authorities in postwar Japan, but also peasant revolt served notice on the British, French and Dutch colonial possessions. Furthermore, by the war's end, Australia and New Zealand had become dependent on the United States rather than Britain for their security, and Britain played no role in the ANZUS Pact of 1951. Responding to the new environment, both strategic and economic considerations influenced the management of Japan's re-entry into the world economy by the US occupying powers. In wartime, military planners reckoned that its future ability to project global power depended on US control of the Pacific Ocean. After the war, confrontation with the Soviet Union in combination with the upsurge of peasant communism and nationalism throughout Asia, demanded a pro-US alliance in the region. In turn, this needed an economic recovery based on a regionally predominant economy, Japan. Indeed, by 1948 Truman and Marshall believed that the future stability and security of the capitalist world economy depended on European and Japanese recovery: 'As the European "containment program" took form, the United States determined that Japan, like Germany, must serve as a bastion against Soviet expansion and, more positively, a catalyst sparking regional recovery'.[34]

Fortunately for such designs, the outbreak of the Korean War settled internal US conflicts in favour of the hawks. As Schaller has demonstrated:

> Japan arose from the ashes of the Second World War largely on the crest of an expanded American military crusade in Asia. . . . [Military expenditures in Southeast Asia] not only helped balance the still chronic dollar gap and compensated for the barriers imposed on trade with China [especially significant given the importance of textiles in Japanese output at that time], but they created for the first time since 1945 an assured foreign market for heavy-industrial and high-technology exports for which no other customer existed.[35]

The peace and military settlements with the United States produced an American–Japanese Co-Prosperity Sphere and the counterrevolutionary stabilization in Taiwan and South Korea provided a stable hinterland for Japanese capitalism, as the United States signed treaties with both states, forming SEATO in 1954. However, despite this external context, Japanese growth in the postwar period was not export-led.[36] Rather, the breaking of trade union power and the linked political defeat and isolation of the socialists enabled the Japanese Ministry of International Trade and Industry (MITI) to orchestrate a sustained policy of import substitution for the large internal market, engineering this through a combination of adminstrative guidance and control over foreign exchange. Together with regional containment (in South Korea and later Vietnam), this allowed 'Japan, Korea and Taiwan [to] move fluidly through a classic product cycle industrialization . . . [as] Taiwan and Korea have historically been receptacles for declining Japanese industries.'[37]

In this context, the state sponsored reorganization of the 1950s brought about a strong development of consumer goods industries. By the 1960s the growth of basic industries meant that manufacturing output had shifted back to the dominance of heavy branches, such that 'the machine tool (including weapons) and electrical machinery and equipment industries reached a peak in 1970. The proportion made up by metal products and oil and coal products reached their peaks in 1973 and 1975 respectively'.[38] A new phase of economic development, which commenced in the late 1960s, deepened in the wake of the restructuring consequent upon the oil shocks of the 1970s, placing emphasis on high-technology, knowledge-based sectors.

Energy and Oil since the War

In the pre-war period Japanese domestic oil production declined from over one-third to less than one-tenth of the refined products market, and imports (80 per cent from the United States) covered the rest.[39] Still, by 1940 oil accounted for less than 10 per cent of Japan's energy needs. Oil supplies were crucial, nevertheless, as the navy was an oil-burning fleet. For this reason alone, the mobilization in preparation for war resulted in extensive state intervention in the industry. Indeed, by the end of the Second World War, the state oil company, Imperial Oil, 'controlled 98 per cent of all domestic fields and many of the overseas fields seized by the military. . . . [But] its parent firms had refused to consolidate downstream'.[40] The refining industry in fact remained divided between eight large companies. More significant still was the destruction of some 85 per cent of refining and storage capacity during the carnage wrought by the US bombing campaign. This was the immediate context within which the US occupying authorities planned the reconstruction of the oil sector. Two dynamics coalesced in this process: on the one hand, while Japanese economic recovery was to be promoted US planners wanted to retain a high degree of political control; and on the other, the US majors wanted to re-enter and dominate Japanese energy markets.

Accordingly, Kennan argued that Japan had to be made economically strong yet politically dependent and suggested that the answer lay in 'controls . . . foolproof enough and cleverly enough exercised really to have power over what Japan imports in the way of oil and other things. . . . [thus giving the United States] veto power over what she does'.[41] Meantime, the oil industry executives directly involved in advising the planners wanted to export refined products from the United States to Japan, so restrictions were placed on domestic refinery construction and repair. But before long a combination of the shift of the United States to the status of a net oil importer (1948), rapidly expanding and diversifying Japanese demand and a shortage of foreign exchange resulted in the construction of local refineries. In the process, Japanese refiners established joint ventures with the majors: equity participation was exchanged for secure crude supplies, and development capital was traded for refinery and distribution facilities. As these deals were cemented, so the US authorities removed controls on the domestic oil industry. Samuels's authoritative study summarizes the position:

> The foreign majors had a rare foothold in a growth market, and a captive one at that. They also had a place to divert Middle East oil when American import quotas were exhausted. The Japanese benefited from stable supplies of crude, capital for reconstruction of their refining industry, new technology, and the prospect of secure access to one of the cheapest factors of production ever available.[42]

At this point, the war damage meant that oil provided a mere 7 per cent of total energy requirements (less than firewood and charcoal) but the rapid economic growth of the 1950s and sixties, based on energy-intensive industries, implied a huge derived demand for petroleum: by 1956 oil was meeting 20 per cent of energy demand and between 1961 and 1973 oil's share in the total increased from some two-fifths to nearly 80 per cent, with around three-quarters of this consumed by industry. Imports met virtually the whole of this expansion and the majors controlled over four-fifths of this trade, sourcing 90 per cent of Japanese needs from the Middle East. Faced with dependence of this magnitude, in policy terms MITI was internally divided and at odds with the Ministry of Finance. Three proposals to create a state company on the European model – the Petroleum Supply Stabilization Fund

(1963), the Overseas Crude Oil Promotion and Finance Agency (1964), and the Crude Oil Public Corporation (1965) – foundered due to opposition from heavy industry and the Ministry of Finance. On the one hand, the position of the majors precluded the creation of an integrated national oil company. And on the other, heavy industry, the electricity utilities and other large consumers wanted to maintain the cheap oil provided by the lax international oil market of the 1950s and sixties, thereby 'opting for economy over security'.[43] The only state venture of significance, the Japan Petroleum Development Corporation (JPDC), was formed in 1967 and was to play the role of a bank rather than that of a national oil company: its purpose was merely to encourage and finance private-sector, overseas investment in oil production.

Responses to the Oil Crisis

By 1973–4, then, Japan was almost completely dependent for its energy requirements on Middle East crude controlled by the majors. Another shared interest linking Japan and the United States was their place as the largest consumers of oil and oil products in the world: in the late 1970s, the United States and Japan together consumed nearly one-half of the non-communist world's oil production and two-fifths of its imports.[44] When the Arab oil embargo began Kissinger urged the Ministry of Foreign Affairs – which had become 'increasingly dependent on the United States Department of State for information as well as policy guidance'[45] – to stand firm. But, under pressure from the Federation of Economic Organizations, the Keidanren, the Tanaka cabinet endorsed the position of the Organization of Arab Petroleum Exporting Countries (OAPEC) in November 1973, marking 'the first open break with American foreign policy in post-war diplomatic history that Japan had dared to make'.[46] Moreover, pressed into action by concerns over the security of oil supplies, Japan's principal forms of adjustment to higher energy prices sought to diversify its oil sources and encourage direct deals (involving itself in non-OPEC production and marketing ventures), to open and expand trade with China, and to restructure its domestic economy, therefore changing the pattern of foreign economic policy.

Diversification of sources and direct deals were closely related and progressed steadily: by 1976 economic agreements had been signed with Iraq, Saudi Arabia, Iran and Qatar; by 1978 direct deal crude oil imports, arranged either through governments or between Japanese importers and state exporters, came to 19.4 per cent of Japan's total crude oil imports; and by 1980 (after the second round of major price increases) direct deals accounted for 45 per cent of imports. (The majors also then carried about 45 per cent of imports).[47] In 1978 the JPDC, now renamed the Japan National Oil Corporation, in addition to building a national stockpile of crude oil, financed some two-thirds of all Japanese overseas exploration. And by the mid-1980s, such state-supported ventures represented around 75 per cent of Japanese-controlled imports.[48] On the domestic front, the shift away from oil was as dramatic as it was urgent; and it involved both an overall improvement in energy efficiency and a decline in the share of oil in total energy requirements. Between 1965 and 1970 Japan's energy demand elasticity to gross domestic product averaged 1.7, whereas it was only 0.2 over the period 1973–80. This resulted from the profit-seeking behaviour by large-scale industry, combined with MITI's administrative guidance, which steered the economy towards: a general decline in the share of (energy-intensive) manufacturing in total output; energy conservation in manufacturing, together with a move away from raw and intermediate materials processing industries into high value-added, less energy-intensive products; an improved energy efficiency of consumer durables; and a gradual rise in energy efficiency in both the transport sector and the home.[49] As a result, oil's share in total requirements fell from 77.6 per cent in 1973 to 61.6 per cent in 1983, and a predicted 50 per cent by 1995.

Despite these efforts, Japan's dependence on imports for energy remains considerable. Thus gas consumption, which is growing rapidly, is provided for mainly by imports of liquefied natural gas (LNG), primarily coming from Indonesia (which supplies around 50 per cent of the market) and Malaysia, but also from Alaska, and in the future from Australia and the offshore deposits around Sakhalin Island. (Japanese power utilities, gas corporations and steel firms account for some 80 per cent of Indonesia's LNG output.) For its coal imports (some four-fifths of consumption) Japan relies upon Australia and North America but also on South Africa and China. Finally, there is nuclear power. In the decade following the 'oil crisis' the proportion of electricity generated by oil fell from 67 to 37 per cent, the contribution made by LNG rose from 2 to 14 per cent, and the nuclear component increased from 2 to 21 per cent. Since it is used for baseload capacity, nuclear power will supply around 35 per cent of electricity generated in the mid-1990s (though being only 22 per cent of installed capacity). And the IEA has estimated that one-half of the new nuclear capacity coming onstream in its member states by 2000 will be in Japan. To secure the future integrity of nuclear power, Japan has been closing the loop of its nuclear fuel cycle by a combination of participation in foreign uranium exploration projects, building reprocessing facilities and promoting fast breeder reactor programmes.

The Opening to China

In a further attempt to expand the sources of its energy requirements, Japan has promoted links with China. As far as the energy sector is concerned, China's growing integration into the world market is primissed on two complementary processes. First, a potentially energy abundant country (rich in coal, but also in hydropower and possibly offshore petroleum) set on a strategy of rapid modernization, since 1976 China has become increasingly dependent upon the West for the capital and technology required to exploit these resources. For example, the World Bank's investigation of China's development prospects in 1981 (which in its recommendations paralleled elements of government thinking) paid close attention to 'agriculture, energy (petroleum, coal, electricity/hydropower) and its conservation; transportation; trade expansion; and the export role of raw materials (especially oil and coal) and textiles and other light manufactures'.[50] Therefore, secondly, China is both expected and will need to become an exporter of energy and minerals in order to provide the foreign exchange necessary to finance its domestic reforms. This pattern of insertion into the world economy, confirmed by the reforms launched in 1978, will be reinforced by China's relative inability to compete in other markets: for while textile exports have improved markedly in the 1980s, the Asian NICs pose an acute competitive challenge across a broad range of consumer goods industries.

China's foreign policy in the 1970s appeared to complement the economic strategy by engineering a 'strategic realignment with the West as the underpinning of a modernisation strategy that aimed at gaining access to Japanese and Western civilian and defence technology'.[51] Thus, parallel to the United States–China *rapprochement*, in 1972 Japan and China normalized relations and the Keidanren thereafter sought to expand bilateral trade on the basis of an exchange of Japanese steel for Chinese oil. Building on this, a long-term trade agreement (LTTA) was concluded in 1978, the logic of which has been ably summarized by Newby:

Clearly, the image of China's natural resources fuelling Japan's modern economy with energy and raw materials, while Japan promoted China's modernisation (and incidentally fostered a vast market for Japanese goods), appealed to both governments . . . For China, the LTTA was an attempt to convince sceptical Japanese of China's commitment to a more open economic policy, while also discouraging the involvement of the Keidanren in the Siberian development project – a matter of considerable concern to Beijing for both economic and strategic reasons.[52]

On the Japanese side, some projections in the 1970s suggested that anything from 35 to 50 per cent of crude oil imports might come from China by the end of the 1980s. In fact, a combination of circumstances rendered prognoses of this kind utterly redundant. Within China, technical difficulties, a lack of economic development and a reassessment of strategy (which curtailed plans for large-scale exports until domestic energy needs had been determined) allowed little scope for export growth. Equally, in the lax oil market of the second half of the 1980s, Japan has been reluctant to significantly expand China's share of its oil imports (3.8 per cent in 1980 to 7.1 per cent in 1986); and has not wanted to exacerbate the trade difficulties with its existing coal suppliers, Canada, Australia and the United States. Thus, while 'Japan is China's principal export market for both coal (taking 46% of exports in 1985) and crude oil (taking nearly 33% in 1985)', these exports 'account for only about 5% of Japan's total coal and crude oil imports'.[53] Nevertheless, prompted by the Keidanren and MITI, Japanese capital maintains a steady if low-key interest in China's energy markets; and over the longer term the complementarity between Japanese energy requirements and Chin's programme of modernization could indeed increase substantially.[54]

Adjustment through Economic Policy

Overall, however, within the highly energy-dependent Pacific region, the position of Japan remains very exposed to energy imports. While significant cuts in the share of oil in total energy requirements have been made, the anticipated switch towards imports from China proved elusive. Indeed, by the mid-1980s Japan still used oil for some 60 per cent of its total energy needs, of which perhaps three-quarters came from OPEC sources. Accordingly, Japan has been constrained to rely on a large-scale increase in nuclear generated electricity. In addition, compensatory changes in economic policy have facilitated adjustment to higher energy prices. In fact, both the opening of new sites of offshore capital accumulation in energy (whether in the Middle East, China or Australia) and the development of technical- and knowledge-intensive industries with low energy inputs have contributed to the liberalization of Tokyo's capital and money markets. By opening domestic markets and encouraging capital exports, Japan's adjustment to higher energy prices has been a part of, and a complement to, its structural adaptation to its current account surpluses.

Postwar Japanese growth was, as noted, heavily based on a bloc of energy-intensive, heavy industries, while coal, textile and swathes of agricultural production were scapped. In turn, this mode of accumulation was underwirtten by a strategic alliance with the United States and the related supply of cheap Middle East oil by the majors. The 'oil shocks' argued for and encouraged a rapid rundown of the basic materials industries as well as a corresponding increase in machine industries. (The rise in energy prices increased the share of mineral fuel in the total value of Japanese imports from 20.7 per cent in 1970 to 50.6 per cent by 1981.)[55] The adjustment was dramtic: as the share of exports in gross domestic product rose from 9.5 per cent in 1970 to 13.3 per cent in 1985, the commodity composition of trade altered radically; in 1970 chemicals, metals and metal products together accounted for 26.1 per cent of Japanese exports and machinery and equipment for 46.3 per cent but by 1985 the proportions were 14.9 and 71.8 per cent, respectively.[56] Greatly encouraged by the appreciating yen after 1985, restructuring of this kind, in association with the relocation of basic industries to South Korea and Taiwan, enabled Japanese capital to 'retain its dominant position in the international division of labour it envisages in the Pacific region by focussing more on the production of capital and luxury goods, leaving mass consumer production increasingly to its subsidiaries in Asia'.[57]

In particular, the combination of a number of domestic factors – a high savings ratio deriving from low levels of socialized welfare provision, reduced domestic investment as postwar infrastructural projects dried up, and continued rapid productivity growth – and the reshaping of the commodity composition of Japan's trade produced a growing surplus in the balance of trade in the 1980s, further augmented by the fall in energy prices after 1985. Now, while trade within East Asia should surpass North American–Pacific trade in the early 1990s, and while Japan already does more trade with the Asian NICs (South Korea, Taiwan, Hong Kong and Singapore) and the remaining Association of South East Asian Nations (ASEAN) economies (Indonesia, Malaysia, Philippines, Thailand and Brunei) than with the United States, 'it is the relationship with America, not with any of its Pacific neighbours, that remains Japan's most important concern, and will continue to do so'.[58] This is so for a number of reasons. If the patterns of restructuring sketched above, further encouraged by fears arising from the European Community's programme for 1992 and the 1987 free-trade agreement between the United States and Canada, are prompting Japan to promote regional integration, then the medium-term erosion of the trade surplus and rising US imports will reinforce the centrality of the United States, as will Japan's concern to retain access to the huge US domestic market. Further, if investment patterns are also considered, then the US–Japanese balance favours the former; thus, prior to the yen revaluation of 1985,

> in 1984 Japan imported $25.6 billion worth of goods from the United States, and $43.9 billion worth were produced by American firms located in Japan, for a total of $69.5 billion. Japan exported $56.8 billion worth of goods to the United States, and $12.8 billion worth were produced by Japanese firms in the United States, for a total of $69.6 billion.[59]

And finally, the bilateral security relationship remains central to Japan's foreign economic policy and strategic alignments, evidenced by growing defence co-operation as the Japanese naval buildup and increased air surveillance supplements US power in Asia, while yen surpluses came to play a central role in financing the US deficits. For it remains in Japan's long-term interests to continue to underwrite the superpower status and role of the United States.[60]

The Soviet Union: Petroleum in a Planned Economy

Introduction

The story of Soviet petroleum provides a dramatic contrast with the role of oil in the capitalist world and in the organization of US hegemony. In fact, it illustrates that control over a crucial energy resource does not in itself result in the ability to project economic and strategic power across the world economy and throughout the state system. Power cannot be explained by control over resources because it is not a property which is reducible to the possession of material (or symbolic) means by agents or institutions. In any full social analysis, it is apparent that power is also a relational property pertaining to the position of agents and institutions within broader social structures. Thus, control over such a vital resource as oil takes very different forms, and has very different implications for enterprises and states, in centrally planned and capitalist economic systems. Indeed, by contrast to the West, the Soviet extension of its influence through the oil industry resulted not in the exploitation of others' resources and the excercise of tutelage over allies, but in the strategically motivated subsidization of its clients. Moreover, such relations as were established in the Third World have in general proved tenuous as the attractions of the world market beckoned. In Western Europe and Japan any relations of power and dependence run from West to East, rather than the other way around.

The Early History

Prior to the Bolshevik Revolution, Russian oil production centred on the Baku fields on the Caspian Sea and was controlled by Nobel and Rothschild interests.[61] On this basis, the 1870s and 1880s witnessed a challenge to Standard's dominance of the European market. With the coming of the Revolution, however, these activities were nationalized. Notwithstanding the British attempt to establish an independent state of Azerbaijan and Standard's contract with its government in 1919, Allied intervention during the Civil War failed either to overthrow the Bolshevik state or to reinstate Western (capitalist) property rights. But Standard Oil, Royal Dutch/Shell and the Nobel interests had by 1922 agreed to a boycott of Soviet oil, forming the Front Uni; and while Europe's demand for Russian oil meant that exports continued, this marked the beginning of a long period of hostility between the majors and the international extension of the Soviet oil industry. Nevertheless, again in a pattern which was to be repeated in the period after 1945, the ending of War Communism and the beginning of the New Economic Policy (1921) produced a series of ventures with Western capital which breached the Front Uni's embargo. In turn, between one-quarter and one-third of this restored and new production was set aside for exports to the West, accounting for a maximum of 18 per cent of the Soviet Union's foreign earnings in 1932.

The basic statistics of Soviet oil production during the interwar period can be simply stated. In the first decade of the twentieth century, the Russian oil industry failed to maintain its initial growth rate, and by 1913 output stood at 10.3 million tons (compared with 33.1 million tons in the United States). After the recovery from the First World War, the Revolution and the Civil War, during the course of the first five-year plan (1928–32) oil production doubled to reach 21.4 million tons. In the second plan (1933–7) while output was targeted to reach 46.8 million tons, in fact low growth of the Baku field in conjunction with the slow development of the Ural–Volga region resulted in a figure of only 28.5 million tons. Performance continued to be disappointing in the third five-year plan, and German occupation during the Second

World War reduced Soviet oil production to 19.4 million tons by 1945. Despite this poor performance, the Soviet Union built up significant export markets in both Europe and Asia during the interwar period.

Postwar Oil

Postwar growth, however, exceeded even Stalin's expectations. In fact, the oil industry overfilled its plan targets and output doubled every five years for the first two postwar decades. Moreover, the discovery of new reserves in the late 1950s, together with Khrushchev's desire to expand the chemical industry and transform the coal-based fuel balance, produced a new emphasis on oil and gas production in the 1959–65 plan. (Soviet output exceeded that of the United States in 1975.) Despite this impressive performance, Soviet productivity remained low by international standards. In this respect, plentiful reserves may have diverted attention away from the need to improve efficiency, for as Goldman has pointed out: 'The output in west Siberia seemed to compensate for the drop in productivity in the Volga–Ural fields in the late 1960s and early 1970s, just as the coming on line of output in the Volga–Ural regions in the late 1940s and early 1950s offset the declining output of Baku'.[62] Indeed, the conjunction of the absence of incentives for increased quality in the Soviet planning system, the treatment of raw materials as a free good (at least in part), and the general failure of administered prices to reflect costs of production has produced an industry which, although a world leader by scale, lags seriously behind its Western competitors in technological efficiency.

None the less, this growth, along with the accompanying expansion of technical expertise, provided the basis for an attempt to use oil exports as an instrument of foreign policy. Immediately after the war, and until 1954, the Soviet Union was a net importer of crude oil and refined products, with Romania providing 80–90 per cent of supplies. As domestic production and refining capacity rose, a significant export potential was generated. From the mid-1950s onwards Soviet oil exports to Eastern Europe (except Romania) increased, typically providing '80 per cent and [during the 1970s] 90 per cent of all East European [oil] imports'.[63] In addition, prior to the Sino-Soviet split, the Soviet Union exported considerable quantities of refined products to China, nearly one-quarter of all Soviet refined products in 1955. Elsewhere, beginning with such neutral states as Finland and Sweden, but shortly followed by NATO members such as Italy and Germany, the Soviet Union sought to expand exports to the capitalist bloc. In this context, given

the determination of ENI under Mattei's leadership to break the closed
circuits of the majors, the case of Italy provided a case of acute concern.
By 1962–3 ENI was importing nearly two-fifths of its oil needs from
the Soviet Union through a barter arrangement which priced Soviet
deliveries appreciably below that of the Middle East supplies.

Meantime, Soviet penetration of the Third World increased. As the
power of the erstwhile European colonial powers declined throughout
the Third World, the Soviet Union provided developing countries with
oil exports (as in the cases of Cuba, India and Guinea) and with both
financial and technical assistance in the oil sector (as in Iraq and Syria).
Once again, not only was Soviet oil offered at prices below those set
by the majors, but also payment could take the form of either soft
currency or barter trade. And in those cases where the Soviets were
providing technical and financial assistance, this was ofen paid for by
oil imports. (Similar considerations applied to weapons sales.) In
another move during the mid-sixties, designed to free Soviet oil for
hard-currency exports, the Soviet Union pressed its East European
allies to increase their trade with Third World states which could pay
in oil and minerals. Now, while national oil companies often regarded
a Soviet presence as a useful bargaining card *vis-à-vis* the majors, the
oil companies regarded such moves as an attempt 'to encourage state
control of oil in free world countries and to incite the leaders of the
developing nations against private oil industry'.[64] Thus, as perceived
by the majors, Soviet initiatives were designed to undermine the Western oil order and with it the free world, and as reckoned by the
United States several NATO partners (especially Italy) were becoming
dangerously dependent on Soviet oil.

Together, this threatened to damage the control of the majors over
reserves and production (particularly in radical nationalist regimes)
while the United States was in danger of losing some of the power it
derived from Western Europe's need for Middle East oil. In response,
'the oil companies appealed to Congress and NATO to support them
in their efforts to shut off Soviet exports';[65] and the early 1960s witnessed US pressure applied through NATO channels to embargo petroleum pipeline components to the Soviet Union the imposition of a
ceiling on Soviet energy exports to Western Europe, and an attempt
by the World Bank and the IMF to persuade developing countries (for
example, India) to abjure barter deals with theEastern bloc.[66] But with
the advent of *détente*, the early 1970s saw Soviet technical assistance
rendered to, and purchassing agreements signed with, Iraq and Libya.
And once again, the Soviet Union also sought and was offered Japanese

TABLE 5.6 Soviet Oil Production, 1945–1988 (million metric tonnes)

	1945	1950	1955	1960	1965	1970	1975	1980	1988
Output	19.4	37.9	70.8	147.9	242.9	353.0	489.3	604.1	624.0

See Fiona Venn, *Oil Diplomacy in the Twentieth Century*, Macmillan, London 1986, pp. 171–77; and *BP Statistical Review of World Energy*, London 1989.

and West European capital and technology in exchange for oil and gas exports.

Throughout this period, it is evident that coal has played a larger role in the Soviet and East European economies than in either Western Europe or the United States. Moreover, in marked contrast to the situation in Western Europe, the coal industry in the Eastern block (as in the United States and China) expanded after 1945. Thus the large expansion of petroleum production and use did not take place at the expense of coal. In part, this was due to the rapid, energy-intensive (indeed energy-inefficient) heavy industrialization of the postwar period which provided a hugh derived demand for all fuels. Yet in addition, the promotion of oil and gas exports to its Comecon partners in order to forestall any turn towards the world market, together with the sending of petroleum to Western Europe to earn foreign exchange, have increased the share of coal in the Soviet Union's energy balance.[67] Indeed, by the 1970s the Soviet Union was exporting nearly one-quarter of its total oil production, with a little more than one-half of this destined for its partners in the Council for Mutual Economic Assistance (CMEA). The opportunity cost of this was huge: if oil and gas exports to the West earned $14 billion in 1980, perhaps $18 billion was foregone by sales to the Eastern bloc.[68] Furthermore, the East European states paid for Soviet oil by exports of manufactures at inflated prices which could not be sold on Western markets. But since the Soviet allies in the Middle East and North Africa have little oil, and given that Eastern Europe has neither the export earnings nor the financial capacity to cover hard-currency oil imports, the alternative to the Soviet supplies would have to be reduced growth in the short to medium term and a major realignment of their economies towards the West over the longer term.

Overall, the position of the Soviet Union's oil production, its oil and gas balance in recent years and its oil exports in the postwar era can be seen in tables 5.6–5.8.

TABLE 5.7   Soviet oil and gas production

	Oil	Gas
1978	572.5	324.3
1988	624.0	693.7

Source: BP Statistical Review of World Energy, pp. 4 and 22.

TABLE 5.8   Soviet oil exports (million tons) 1956–1988

1956	1966	1976	1988
4.8	71.6	141.3	219.0

Source: BP Statistical Review of World Energy, p. 18; and Goldman, The Enigma of Soviet Petroleum, pp. 74–5, table 4.3.

In 1985 Mikhail Gorbachev declared that 'the epoch . . . of "easy oil" . . . is coming to an end',[69] and he might have added much the same for coal. The Soviet Union now faces a series of acute dilemmas in its energy planning, and at the heart of these are the problems of nuclear power and Siberian development. In the context of declining productivity in the European coal industry and growing difficulties in the oil sector (though after a period of stagnation in the mid-1980s production increased again in the late eighties), and given a general desire to conserve petroleum (increasingly gas) for hard-currency exports, Soviet planners have decided to increase the role of electricity in final energy consumption and raise the share of this generated by nuclear power.[70] Specifically, exports of nuclear-generated electricity to Eastern Europe along with nuclear power station construction in the region are set to increase. And in this programme, which is coordinated through the CMEA, the 'Soviet Ukraine has been assigned the key role in East European nuclear development'.[71] Indeed, despite the intervention of the Chernobyl disaster (April 1986) between the Twenty-Seventh Party Congress (March) and the presentation of the Twelfth Five-Year Plan for 1986–90 to the Supreme Soviet in June, Premier Ryzhkov announced that the 'growth of atomic energy' would play an increasing part in electricity production.[72] However, the rapidly growing political independence of many East European states from the

Soviet Union and their growing ties with the European Community allied to the spread of nationalist protest throughout the Baltic regions and the Ukraine, poses acute obstacles to such a strategy.

To compound the situation, the problems associated with the exploitation and mobilization of Siberia's huge resources are of central concern to the Soviet state. Flanked by the buoyant Japanese economy and a rapidly modernizing China, Siberia is a thinly populated, resource-rich domain, with little infrastructural provision and only limited and vulnerable communication and transport links with the European areas.[73] Therefore, the administrative integrity of the region has become increasingly salient to Soviet military strategy and planning, as evidenced in the establishment of a separate theatre of war in the Soviet Far East in 1979. Perhaps of more immediate importance, however, is the immense economic potential of the region, 'since it contains almost three-quarters of the country's mineral, fuel and energy resources, over half its hydro-electric resources, about half of its commerical timber reserves, and one-fifth of its cultivable land'.[74] In the case of energy resources, this geography is part of a more general imbalance in which around 90 per cent of energy use is west of the Urals yet 90 per cent of reserves lie to the east.[75]

The scale of the difficulties facing the Soviet economy in this sphere are rarely appreciated in the West and should be underscored. First, the primary and extractive industries now account for around one-half of the national product and total investment, currently even expanding in relation to the manufacturing and service sectors as rising proportions of investment are devoted to opening up new areas of the country. Although current investment policy is in principle geared towards securing 75–80 per cent of fuel and raw material consumption increases from improved efficiency, thereby stabilizing the share of total resources taken by these sectors, growing economic and geographical problems have meant that more that 90 per cent of new capital investment in mining is needed simply to maintain existing levels of output.[76] Thus, of total planned investment for 1986–90 nearly one-fifth is still directed to the energy complex.[77] Second, the social and political challenges thus created are overlain by the demographic-cum-national problems which appear to have taken a decisive turn for the worse in the late 1980s. The declining birth rates in the Russian, Ukrainian, Latvian and Estonian republics, which currently account for some 80 per cent of the labour force, in conjunction with rising rates among Turks, Uzbeks and Taszhiks pose 'an economic, and not simply a nationality problem . . . [because] the declining areas possess 85 per cent of the country's industry, 82 per cent of its electricity, and equivalent proportions of related infrastructure. . . . [Moreover] the new sources of energy lie in regions devoid of infrastructure *and* population'.[78]

The one bright spot in the Soviet energy balance is the potential future of natural gas. By 1988 the Soviet Union produced 40 per cent of the world's total natural gas output, held an equivalent proportion of global reserves (44.2 trillion cubic metres); and production has more than doubled (to surpass Soviet oil output) over the last decade. Thus, although the Soviet Union had a reserves to production ratio for oil of only 13 years in 1988, the current reserves to production ratio for natural gas is 55 years.[79] On top of this, in August 1989 the Soviets announced the existence of two new huge offshore gas fields in the Barents and Kara seas, perhaps equivalent to more than the total current reserves of North America (8 trillion cubic metres).

Already the Soviets have encouraged deals with Western capital to exploit oil and especially gas reserves, predominantly West European and Japanese capital thus far: typically, Western development capital and technological resources have been exchangd for guaranteed exports of Soviet petroleum. But with the advent of Gorbachev and the econ- omic reforms announced during 1986, culminating in the Supreme Soviet law on joint ventures of January 1987, the prospects of foreign capital developing Soviet oil and gas rsources for profit now loom large.[80] For example, in the case of the Shtokmanovskaya field in the Barents Sea, Finnish and Norwegain capitals are seeking to participate in its development through joint ventures and compensatory trade. The Soviet Union, for its part, is reported to be seeking to acquire the requisite offshore technology from such Western involvement, thus enhancing its capabilities in gas recovery techniques. By these means, the Soviet economy will benefit from the resources developed, the foreign exchange earned and the easing of the energy constraint on Comecon. Further negotiations, involving unprecedented access to commercially sensitive geological data, have taken place with regard to joint ventures with Western oil capital for new onshore oil fields in Siberia, the Arctic Circle and the Caspian region.

The Soviet Union in the Middle East

The Russians were unable to comprehend the dominating role of national- ism in the Arab world.

Mohammed Heikal[81]

Houari Boumedienne to the Soviet leadership after the 1967 Arab–Israeli War: 'The worldwide national revolution is receving successive blows

from American neocolonialism and your friends feel that the slogan of
peaceful coexistence has turned into fetters restricting your movements.
We sincerely wish to know where the dividing line lies'
    The Soviet reply: 'What is your view of nuclear war?

Houari Boumedienne[82]

Khrushchev's accession to power became a turning point in the development of the Soviet oil industry, whose character now changed from
primarily a domestic-oriented industry into a tool of foreign policy . . .
Oil became the major weapon of the Soviet foreign trade offensive.

Klebanoff, *Middle East Oil and U.S. Foreign Policy*[83]

All too often, commonsense Western assessments of the Soviet role in
the Middle East are structured by the putative identification of a
number of distinct motivations behind its regional foreign policy. From
a realist perspective it is argued that the geopolitical interests of the
Soviet Union do not differ significantly from those of the Russian
Empire; specifically, the Soviets are driven by a search for 'warm
water ports'. By contrast, accounts which divine a closer link between
domestic and foreign policy draw attention to the problems relating to
the Soviet Muslim population deriving from religious and political
turbulence in the Middle East. And when characterized as the last great
imperialist power, a direct connection is made between the impending
Soviety energy crisis and the aggressive designs of the Soviet Union on
the region's oil. To these particular determinants, the rhetoric of the
1980s added the linked charges of sponsoring the spread of revolution
and terrorism. None of these claims will withstand close inspection.[84]
    In military–strategic terms, the Middle East has a relatively low
priority in Soviet planning; unlike in either the European or the Asian
theatres, in this region the Soviet Union does not have to reckon with
significant hostile forces deployed on its borders. The threat of direct
attack from the Middle East is minimal. What is more, the Indo-
Arabian region does not compose a unified set of problems for Soviet
planners: in the northern tier – Turkey, Iran and Afghanistan – the
Soviets have been flanked by a NATO power (Turkey joined in 1952)
and a long-standing US ally (Iran until 1979); in the Arab Gulf the
Soviet Union has not had to face any immediate challenges; and in the
eastern Mediterranean and the Indian Ocean the Soviets have had to
contend with the forward deployment of US nuclear capabilities. In
each case, however, Soviet aims have been revealed to be similar: first,
'to prevent the emergence of an Iranian U.S. alliance or the use of
Iran as a U.S. base',[85] second, to prevent the West from gaining a
general political and strategic advantage in the region as a whole; and

third to offset the West's naval superiority. Thus, not only are the southern military districts of the Soviet Union in a low state of readiness but also its naval buildup in both the Mediterranean and the Indian Ocean has been directed 'to counter the West's seaborne strike capacity rather than to give Moscow the power to interfere with shipping on the high seas or intervene locally on shore'.[86]

From the standpoint of the Soviet Union, however, the effect of the formation of the Baghdad Pact (1955) and later Cento (1959) was to: 'transform what had been an effective buffer zone in the prewar period into an important link in the worldwide chain of Western containment strategy, but it also meant the extension of NATO's military power to the USSR's backyard, thus turning it into a potential theatre of war'.[87]

Into this context, after Stalin's retreat from northern Iran at the start of the Cold War, the first significant Soviet opening to the Middle East came with Khrushchev's accession to power and the formal revisions to foreign policy adumbrated at the twentieth Party Congress (1956).[88] During this period, optimism that the Soviet Union and its model of socialist development could triumph over the West in a peaceful competition for political and economic influence coincided with the coming to power of a wave of revolutionary and radical nationalist movements across the Third World. From 1956 onwards Egypt provided a model for Soviet policy, as military and economic ties were extended, and during the mid-1960s the Soviet Union encouraged Nasser to expand state control over the economy and urged local communists to join the Arab Socialist Union. Soviet relations with Syria improved after the left wing of the Ba'ath Party assumed power in 1966 and again as Soviet arms became its mainstay after the 1967 Arab–Israeli War. In addition, relations were also established with Iraq and later with Algeria and Libya in North Africa. However, as is well known, the Soviet position in Egypt counted for little when in 1972 Sadat demanded the removal of its military advisers and curtailed access to Egyptian bases. The attempt to compensate for this loss by increased assistance to Syria and Iraq, along with increased support for the Palestine Liberation Organization, could not disguise the magnitude of the Soviet setback in the superpower competition for regional influence. As in Egypt, so in Syria and Iraq, treaties of friendship and co-operation, combined with large-scale military transfers, economic aid and diplomatic support, did not alter the social base of the regimes in question towards a model of socioeconomic development integrated into the socialist bloc.

Indeed, with the ousting of the Soviets from Egypt, Moscow's dependence on Syria increased, But a broader coalition proved altogether harder to fashion, as Iraqi and Syrian enmities remained deep. After

Sadat's *rapprochement* with Israel in November 1977, and given Syria's growing isolation in the Arab world and accumulating internal difficulties, the Soviet influence over its ally expanded. However, the wider Soviet goal of forging a radical Arab bloc – consisting of Syria, Iraq and the Palestine Liberation Organization together with Algeria and Libya – foundered on the perennial rock of regional strategic and political rivalries. Thus, while it was able to obstruct Saudi attempts at reaching a common Arab stance towards the Reagan Plan of 1982, Syria's assistance to the Soviets decreased as the conservative Arab states were angered by its support for Iran in the Gulf War and the deteriorating position in Lebanon. In July 1984 the Soviets resumed diplomatic relations with Egypt and in 1985 Gorbachev established relations with Oman and the United Arab Emirates while stressing the Soviet concern to normalize relations with Egypt. This growing Soviet freedom of manoeuvre towards to conservative Arab states, then, was founded on the failure of a radical coalition to cohere.

In accounting for the failure of the Soviet Union to make lasting and significant gains in the Middle East, emphasis is often placed on the centrality of the Arab–Israeli conflict, in combination with the role played by the United States in providing the military and economic underpinnings to Israel's regional dominance. For in successive conflicts – the Arab–Israeli wars of 1967 and 1973, the Syrian intervention in Lebanon in 1976 or the Israeli invasion of Lebanon in 1982 – the Soviet Union has been unwilling or unable to arm and aid the Arab combatants sufficiently to defeat Israel, urging caution upon its allies and preferring to seek a negotiated settlement to the Palestinian question with itself and the United States as guarantors.[89] But, from Kissinger onwards, the United States has routinely refused to allow meaningful Soviet participation in any multilateral diplomatic process. Accordingly, if the Soviet Union cannot provide the Arabs with a solution to their dispute with Israel, and if their rising incomes can purchase advanced weaponry from the West, what influence can the Soviets bring to bear in the region? This is a cogent argument, but it is not by any means the whole story.

As noted, in an environment provided by an apparently improving correlation of forces, Soviet policy was founded on the belief that the independent states of national democracy could advance from colonial underdevelopment to communism without passing through a capitalist stage, provided assistance was forthcoming from the developed socialist bloc. Khrushchev's oil diplomacy of 1956–64, then, must be seen as an attempt to internationalize Soviet-type productive forms, based upon bureaucratically administered development and state planning.

Moreover, raw materials in general and oil in particular were one area
where the Soviets could compete with the West. (The other principal
domain of competition was arms disbursements.) Thus, by offering cut
rates, help in refinery construction, barter trade, and payment in soft
currencies the Soviets posed a symbolically and politically significant,
if economically limited, challenge to the majors. In the case of Syria,
for example, the Soviets have provided extensive aid to develop infrastructure, agriculture, transportation and industry, 'where the USSR
has helped to establish Syria's oil industry by increasing the state's oil
storage capacity and assisting Syria in setting up its national drilling
company'.[90]

Indeed, while the dominant mechanisms and tendencies operating
against the majors' control lay in the capitalist world and the consolidation of independent states in the periphery, the impact of Soviet
support for Iraq and Libya, the spearheads of OPEC's challenge in the
early 1970s, is not to be underestimated. In general, however, Soviet
policy both involved a profound misreading of the pattern of socioeconomic development in the Middle East and indicated the failure of the
Soviet model of development to provide a long-run, generalizable
alternative to integration into the world market. On the one hand, as
Samir Amin has argued, the nationalization of foregin property by
radical nationalist regimes or the creation of a public economy based
on oil rents produced not aspirant socialist regimes but a series of state
capitalisms which, increasingly after 1967 and 1973, became tied to the
world market, thereby fragmenting Arab unity, undercutting the basis
for a socialist path of development and strengthening the Western
position.[91] On the other, the autarchic pattern of extensive economic
growth in the Soviet bloc – with its limited alienability of assets,
the consequent absence of exchangeability between money and other
resources, and therefore the immense disutility of the rouble as compared with the dollar – meant that the structural power of the Soviet
Union in the capitalist world market, whatever the relative level of
material resources mustered, would be low by comparison with that of
the United States.

Finally, the limited advances made by the Soviet Union in the Middle
East and North Africa, together with the more significant revolutionary
upheavals throughout the Third World in the second half of the 1970s,
contributed a great deal to the US disenchantment with *détente*, provoking the shift to Cold War and Reaganism in the 1980s. This overarching
conflict further circumscribed Soviet entanglements in the region, for,
as Spechler has noted:

The major military buildup, both nuclear and conventional, which the Americans had undertaken after 1979, the establishment of the Rapid Deployment Force and a network of bases to support it, and the pursuit of an anti-Soviet 'strategic consensus' in the Middle East were seen in Moscow as, at least in part, a response to Soviet actions in the Third World.[92]

Thus, to the degree that the Soviet Union's ability to compete for political hegemony in the Third World ultimately derives from the domestic strength of the Soviet economy, and to the extent that the superpower arms race saps the latter's productive potential, to say nothing of the subordination of political and ideological struggle to the dictates of security, the onset of the Second Cold War resulted in a profound reassessment of Soviet foreign policy.

Conclusions

It is undeniable that the postwar period witnessed an overall loss of direct US control over the world oil order, and therefore, to the extent that this provided a key component of US global hegemony, the general ability of the United States to act as world leader has weakened. In the case of Western Europe, once dependent on US-sourced Middle East oil, following the relative eclipse of British and French imperialism in the region, the 1970s and eighties have been characterized by an increasing freedom of movement with respect to both the Eastern bloc and the Middle East. As for Japan, from being more or less accurately seen as a sector rather than a state or a ward under US guardianship, the 'oil crisis' signalled a remarkable expression of independence in respect of policy towards China and the Middle East. But the underlying picture is not as clear as these bare facts suggest. Western Europe and Japan are turning towards economically weak communist powers on the one hand, and to a region whose security is guaranteed by US military power on the other. In addition, the economic power of the United States remains quantitatively and qualitatively superior to that of its capitalist allies and competitors.[93] Finally, burden-sharing notwithstanding, the ultimate security of the European Community and Japan is equally secured by the US nuclear umbrella. Overall, this context involves the United States in a complex set of relationships by virtue of which its directive capacities continue to exceed substantially those of its rivals. In the case of oil, this can be seen by a final look at the economic and strategic context in which Western Europe, Japanese and Soviet decisions are situated and taken. Reversing our order of presentation, let us start with the Soviet Union.

Whatever the ultimate dynamic and purpose of the current programme of reform in the Soviet Union, it is apparent that for the foreseeable future the overriding priority will be internal reform or crisis management, and that foreign policy goals will – in so far as security concerns allow – be structured by this need. Above all, the Soviet Union is seeking to bring about and stabilize a new set of relations with the West in general and the United States in particular. Although this is couched in a somewhat shallow rhetoric of 'interdependence' and 'univeral interests', appearing to involve an untutored conception of the workings of the capitalist world economy,[94] we have seen that the reality is one of a Soviet inability to project power globally or to direct the behaviour of other states. The Soviet Union is only a superpower in the military domain, lacking the capitalist mechanisms of control deployed by the United States.

On top of the qualitative problems noted above, in total the Eastern bloc economies are roughly one-third the size that of their Western competitors, the bloc's foregin trade is only one-twelfth of the world total (both EC and US trade with the Third World are some six to seven times that of Soviet exchanges, and half of the latter are military sales), and the Soviet Union's net foreign investment is minimal.[95] Further, Soviet foreign trade is dominated by raw materials, with petroleum exports alone accounting for over 60 per cent of Soviet hard-currency earnings in the 1980s. In this context, just as Khrushchev's oil diplomacy and attempt to compete with the West on the economic plane proved no match for the attractions of the world market to the capitalist classes of the Third World, so Soviet (and for that matter Chinese) energy trade with Western Europe and Japan poses no threat to the capitalist markets. Indeed, if anything, the dependence runs in the other direction: although Western Europe and Japan have much to gain from Soviet and Chinese energy exports, the urgency of reform for the state socialist bloc accords it little freedom for manoeuvre. Moreover, an attempt by the Soviet Union to extract political concessions from such trade would risk compromising the reduced strategic pressure on the Soviet system which is now a primary concern of all sections of the Soviet leadership. And, of course, the intensity of political and strategic pressure is still ultimately regulated by the United States.

If, then, the terms of political and economic competition between East and West have now shifted decisively in favour of the latter, is not US tutelage thereby undermined to the extent that Western Europe and Japan either reduce their dependence on Middle East oil or increase their trade with the larger energy and oil exporters? Consider the case of Japan and China, perhaps the most significant in its long-term implications. As laid down in 1979 by Prime Minister Ohira, Japan's aid to China was intended to assist in the latter's modernization and opening to the West, to remain in balance with aid to the ASEAN bloc and to exclude military transactions. (By the late eighties Japan provided one-half of all development loans to China and accounted for one-fifth of its foreign trade.) Further, while early relations were based upon the assumed complementarity between China's need for capital and technology and Japan's energy requirements, strategic considerations have increasingly shaped Japanese policy. And, as one recent study concluded, Sino-Japanese relations 'remain closely circumscribed both by their respective relations with the superpowers and by relations between the superpowers'.[96]

Within this strategic quadrilateral, the US–Japanese alliance continues to play a central role. For despite recent increases in Japanese military activity, the extension of the US nuclear guarantee to Japan remains the core of the latter's security. In turn, China both has its own military linkages with Washington and an interest in the maintenance of US–Japanese ties, thus precluding a full-scale Japanese military buildup. Moreover, China fears alienating Japan to the extent that its trade and aid migrate to the Soviet Far East in search of Siberian resources. Dibb has highlighted one important effect of this conjuncture: 'Taken together, China's anti-Soviet posture and the prospect of growing Chinese power, substantial Japanese rearmament, and a more active US military presence worldwide, represent a potentially critical challenge to Soviet power in the 1990s'[97]

Gorbachev's Vladivostok speech in 1986, proclaiming the Soviet Union a Pacific power, was thus a 'guard against the establishment of a Washington–Tokyo–Beijing axis'[98] as well as an attempt to promote economic links with China and especially Japan – and late 1989 even saw signs that the Soviets might be reconsidering their position in relation to the Kuriles. However, although all of China's conditions for improved Sino-Soviet relations have matured – a Soviet withdrawal from Afghanistan, reduced border tension and the end of Vietnam's occupation of Cambodia – its position will remain 'independent, but not equidistant'[99] between the superpowers. In sum, two important conclusions can be drawn: first, Japan's interest in China for energy is likely to remain subordinate to overriding considerations deriving from its general ties to the United States; and second, any Sino-Soviet détente

will be tempered both by their central geopolitical relations with the
United States and by their joint competition for Japanese capital to
develop their raw materials.

Finally, what of Western Europe? In the competition for Middle
East oil, both Western and Eastern Europe are hampered by the
facts that perhaps three-quarters of world fuel trade is non-marketed
(intracompany or long-term contracts) and that oil deals between producer states and the majors have increased.[100] In fact, in Africa and
the Middle East the pattern of their trade is similar: raw materials are
exchanged for manufactures. And while the European Community has
an advantage deriving from the qualitative superiority of its manufactures and military hardware as well as its access to hard currency, it
cannot rival the influence which the United States derives from its
special relationship with Saudi Arabia and the Gulf states. Moreover,
although the United States was unable directly to link Western Europe's
(energy) trade with the East either to security concerns or to Atlantic
trade, the actions of West Germany in the wake of 1973–4 and of the
support of the allies for US manoeuvres in the Gulf in 1987 amply
demonstrated the limits to the European Community's unity of action.
In both cases, what finally secured Western unity under US leadership
was the pacification of intra-Western differences by the common interests of all in the overarching East–West competition.

6

US Strategy in the 1980s: Unilateralism and Reconsolidation

Introduction: Hegemony – Diminished or Reoriented?

In this concluding chapter I try to demonstrate how the general realignment of US hegemony is paralleled in, and thus can be illustrated by,
the changes to its position in the global oil industry. As we have seen,
US control of world oil came to play a key role in underpinning its
international power: but just as with US hegemony in general, so with
oil, its position has been more reoriented than diminished. As the 1980s
wore on, concern and debate extended as to the longevity of US
hegemony both in specific fields – trading position, role of the dollar,
control of world oil and the military balance – and in the directive
capacities of composite US state strategies. The Reagan administration
apparently perceived the causes of the US loss of power in the mistaken
policies and weak will of erstwhile regimes, while domestic critics –
broadly basing themselves on realist premisses – argued that its decline
was altogether more deep-seated and thus harder to reverse. For example, the theory of imperial decline advanced by Gilpin suggested that
all empires exhibit an inherent tendency to decline as a result of the
external burdens of leadership, the internal secular tendencies towards
rising consumption at the expense of productive investment and the
global diffusion of technology.[1] Similar arguments can be found in
Wallerstein's cyclical theory of hegemony or in Kennedy's notion of
'imperial overstretch'.[2]

The US share of world output, exports and military spending declined appreciably during the postwar years, especially in the 1950s and sixties.[3] Further, of the major regions of growth in the 1970s – Comecon, the OPEC states and the NICs – only in the latter did trade with the United States expand faster than with Western Europe and even here the US performance was outpaced by that of Japan. It is, therefore, clear that these analyses identify genuine problems facing the US political economy which Reagan's contradictory project did little to address. Indeed, through attempting simultaneously to compete militarily for superiority over the Soviet Union and to meet the economic competition of Western Europe and, particularly, Japan, the Reagan administrations may have exacerbated the problem in so far as they increased the external burdens of the United States, were unable to control the rising competition for technological leadership across a growing range of sectors, and consequently faced acute dilemmas in managing the trade-offs between investment, military spending and consumption.[4]

Valuable as such quantitative considerations are, a range of qualitative considerations also need to be entered: first, beyond noting the material resources of the United States, it is important to locate its position within the component structures of the international system, wherein it still derives considerable advantages; and second, the complex management of these positions is always imbricated in the broader social conflicts – both within the capitalist system and between this and the state socialist societies – which the United States as the hegemonic power, seeks to order. In this regard, it is important to bear in mind that along with the nuclear arms race the focus of the Second Cold War was the 'arc of crisis' and the Third World more generally.[5] And as Halliday has cogently argued, this social or systemic conflict – what Deutscher termed 'the great contest' – introduced a global, bipolar and systemic character to world politics.[6] Thus, for example, whatever the economic legacy of Reaganomics, the competition induced by the Reagan military buildup imposed huge costs on the Soviet economy, at the very least providing another cogent argument for domestic reform and international retreat on the part of Gorbachev.

In fact, in terms of the components of hegemony identified in chapter 2, US interests remain preponderant in the transnational spread of capital within the world market. The dollar continues to be the major form of global liquidity and store of reserves, not least because access to the US domestic market is a fundamental precondition for competitive success. The position of the US economy within the world market is still marked by a high degree of asymmetry as compared with either Western Europe or Japan and the Asian newly industrialized countries.

In addition, large swathes of the state socialist bloc are currently extending their integration into the world market on terms dictated by the West. On the military plane: while the nuclear strategic balance approximates to parity, the United States is far ahead of the Soviet Union in technological development; the relative decline of US military spending within the West might weaken its tutelage over Western Europe and Japan but it augments its bilateral security relationships elsewhere (such with Saudi Arabia) as compared with those of the Soviet Union; and the growing retreat of the Soviet Union from its global position of the late 1970s, together with the relaxation of tension in Europe from the mid-1980s, has extended the US capacity for regional intervention in the Third World.

Politically, the benefits deriving from the relative success of the United States in the Cold War, in combination with the solidifying of the Japanese–US relationship, have in part offset the costs of the accompanying loss of tutelage. On the one hand, the West under US leadership has been able both to force the Soviet Union into a major project of internal reforms and external retreat and to organize the re-integration of numerous Soviet allies into the world market. On the other, based on an exchange of military protection and access to the US market for Japanese financing of the US deficits and political co-operation, Pacific capitalism is now the most dynamic region of the world economy, pioneering the technological and economic developments of the future. And in the Third World, the crises of the Soviet bloc, in conjunction with the continued (albeit uneven) dyanamism of the West, have altered the terms of social and political struggle strongly in favour of pro-capitalist forces. Together with the absolute superiority of the US capacity for regional intervention, this further privileges the international role of the United States.

Drawing on these general themes, this chapter argues that (1) the Reagan project can be best understood as a composite attempt to fashion a unilateral response to challenges across the domains of US world leadership; (2) its position in the world oil market, although in decline remains strong given the dollar denomination of oil and the high overall level of energy self-sufficiency in North America as a whole as compared with either Western Europe or Japan; (3) the US position in the Middle East – central to its directive abilities in the world oil market – remains strong and irreplaceable in the foreseeable future; (4) OPEC cannot be considered as a cartel and in fact increasingly revolves around the Gulf producers, themselves tightly tied to the West; and (5) viewed from this perspective, it is clearly the United States – notwithstanding the rising role of Europe, Japan, and the Soviet Union in the political economy of international oil – which maintains a pre-dominant directive role in this key domain of hegemony.

The Geopolitical Economy of Reaganism

Unilateralism Overtakes Trilateralism

The composite strategy of *détente*, the Nixon doctrine and the limited moves towards trilateralism faced multiplying problems in the second half of the seventies.[7] In 1972 the expulsion of the Soviets from Egypt gave the United States a relatively free hand in the Middle East, while Africa and Latin America presented few problems and in Western Europe Schmidt abandoned an independent German orientation in the latter half of the 1970s for the pursuit of NATO 'modernization'. The crucial weakness of the Nixon–Kissinger strategy, however, concerned the failure of linkage in the second half of the 1970s and the inability of the Carter administration to avert domestic economic stagflation. During the era of *détente*, opposition to bilateral agreements persisted in both blocs, deriving particular support from the military and conservative elements. In the case of the Soviet Union, there was little real improvement in the performance of the economy (as opposed to the benefits of raw material exports, especially petroleum), and in the United States no significant capitalist or electoral interests emerged to support *détente* while strategic parity facilitated Soviet intervention in the Third World. In fact, it was in Western Europe (above all in West Germany) that the advantages of and interest in *détente* were felt most keenly. Within the United States elite opinion had begun to polarize by the time of the 1976 presidential elections: on the Republican right the Committee on the Present Danger and its related forces began to organize against the East coast, liberal bloc where a more Atlanticist and conciliatory attitude prevailed.

Carter's failure to overcome the recession, and more importantly developments in the Third World, including those in Ethiopia/Somalia, Afghanistan and Iran followed by North Yemen, Grenada and Nicaragua,

> combined to produce a shift to the right in US politics from the middle of 1978 onwards, in both popular sentiment and Administration policy. At the same time, the Administration felt its own position strengthened by the progress in developing a closer relationship with Peking. . . . The onset [of the Cold War] took a double form: a resolution of the inherent ambiguity of the Carter programme in favour of Cold War components;

and a larger wave of Cold War mobilization which swamped Carter altogether in 1980 and so swept Reagan to power in January 1981.[8]

The Reagan programme itself comprised a number of elements in an overall contradictory formation, centred on a right-wing mobilization around Cold War liberalism. The domestic roots of this project lay in Reaganomics, which was both consequence and cause of large-scale shifts in class forces, driving the US economy towards new patterns of accumulation with pronounced international ramifications (clearly illustrated by the debt crisis on the one hand, and in reduced policy consultation with the European allies on the other). Politically and ideologically consonant, if economically in some tension with this, was the massive anti-Soviet arms buildup which underpinned the strategy of renewed Cold War. This involved the pursuit of stalemate and confrontation in arms 'negotiations', a moralistic anti-communist crusade, and economic competition with the Soviet Union prosecuted through the arms race and trade restrictions. In turn, both the Cold War project and the unilateral economic liberalism were also directed towards conflicts in the Third World, where a rigid East–West grid of confrontation was imposed on what were often primarily indigenous struggles. The overall aim here was to weaken and if possible overrun the post-revolutionary regimes that had come to power in the 1974–80 revolutionary wave.

The United States and the World Economy

In an important analysis, Davis identified two principal sources of dynamism in postwar US growth.[9] To begin with, Fordist (or Marshallist) relations of social stability and economic growth were spread across metropolitan capitalism, thus re-unifying the world market under the dominance of US transnational corporations. Within the domestic market, captial accumulation deepened through expanding high-wage, primary-labour-market jobs, and it widened as a result of the tax-led formation of the Sunbelt. Davis also showed how the crisis of this pattern of growth produced a demobilization of the working class throughout the 1970s, while the concurrent tertiary and quarternary growth in the Sunbelt was harnessed to the political and distributional advantage of the middle strata and small and medium-sized capital. Together, this led to the formation of a political bloc mobilized around the policies of deregulation, tax breaks, redistribution from poor to rich and the curtailment of Great Society welfare programmes. Combined with the tight monetary policy and high interest rates of the

Volcker period at the Federal Reserve Board, these policies shifted social demand to those with a higher marginal propensity to save, thereby strengthening rentier interests in the profit distribution process. In this way, 'Reagonomics without the Laffer curve [was] a program to reduce consumption to finance investment and military spending'.[10]

So, in the first phase of the Reagan period, the administration followed a tight monetary and high interest rate policy, with the aim of revaluing the dollar in order to control inflation. (Some 40–50 per cent of the decline in the rate of inflation between 1980 and 1984 was due to revaluation.) A predictable if unintended consequence of this policy, with adverse consequences for US export markets above all in Latin America, was the ensuing stagnation in the periphery. By mid-1982, therefore, there was genuine concern within the central banks of the advanced capitalist countries that the international debt economy might provoke a financial crisis with unthinkable results. And added to the deteriorating state of the real economy during the domestic recession, evidenced in the dramatic negative shift in the trade balance, this prompted a shift in strategy. In the summer of 1982 (and again in 1983) the Federal Reserve Board relaxed its monetary squeeze by reducing interest rates, so that the fiscal stimulus due to the combination of tax cuts and increased military spending produced what Lipietz rightly terms a 'text-book Keynesian' recovery.[11] The resulting boom also conveniently renewed the political support of the New Right by strengthening the neo-rentier bloc. Equally, the reorientation of US industry from the Northeast to the Southwest and the global relocation of trade and capital flows continued apace.[12]

The high real interest rates during the Reagan era were also sustained by the decision to raise military spending while cutting taxes. Even without this deficitary militarism, the federal deficit would have risen sharply after 1981 'because of inflation-swollen entitlements, inflation-boosted interest rates, and post-inflation reduced tax revenues'.[13] Above all, by the mid-1980s, a vicious circle had set in as the rapidly expanding interest payments on the national debt became the 'fastest growing major item in the budget'.[14] In turn, high interest rates drove the dollar upwards on the foreign exchanges, with significant adverse consequences for the performance of US exports and the maintenance of home markets against foreign competition. Moreover, as and when domestic manufacturing approached capacity, high military spending would pre-empt resources, producing a mixture of capacity bottlenecks, trade deficits and inflation. Nevertheless, the lynchpin of what Davis characterized as a tendential 'overconsumptionist' mode of accumulation during this period was the warfare Keynesianism of the Pentagon:

'Deficitary militarism may seem to be out of control, but for the present it supplies the necessary cohesion of overconsumption'.[15] This 'militarization of American capitalism' served two further goals. First, the massive military buildup operated as a form of substantive support for the advanced sectors of the economy, especially electronics. And second, as Parboni argued, these military outlays represented 'the last hope for [the US's] ruling groups to recover world industrial leadership, and to bring the reluctant European partners back into the US orbit by subordinating them on the military, industrial and commercial planes'.[16]

On the international plane, a major factor behind the shift of US growth patterns into a overconsumptionist mould was the strengthening of bank, oil and rentier incomes in the metropolitan core as a consequence of the oil price rises. The recycling of the OPEC surpluses greatly augmented the burgeoning stateless Euromarkets, while the second oil shock was translated – through the agency of monetarism – into high interest rates in the core and stagnation in the periphery. Encouraged by this, manufacturing and transporation capital flowed into the new high profit sectors of energy resources, financial services, real estate, emergent tcchnologies, and of course defence. This further undermined the old Fordist indusdtries and strenghthened the position of financial conglomerates oriented towards the Sunbelt. The two rounds of price increases also meant that both the majors and the independents received windfall profits. The political corollary of these shifts has been charted by Edsall: the independent oil political action committees were central in ending Democratic control of the Senate and provided key backing and support for the election of Reagan, engineering a shift in the locus of control of the Republican Party.[17] Davis has commented on the gcncral import of this:

> the oil independents are the core of a new power-bloc which, thanks to the continuing shift of capital and tax revenues to the West and the South, is displacing Northeastern multinationals in the active control of the Republican apparatus. In this sense, the recent near-extinction of 'moderate republicanism' – i.e., the Dewey–Rockefeller wing dominant from 1940–64 – is part of a larger pattern involving the supplantation of Fordism and the rise of new rentier and military–contractor networks.[18]

Globally, the complement to these developments was a marked changc in the pattern of metropolitan trade. During the 1970s, the US manufacturing deficit with East Asia was balanced by a surplus with Western Europe; the US trade surplus with Latin America was offset by capital exports to the region; the West European deficit in manufacturing with the United States and Japan was balanced by the export of capital goods to the Third World (particularly Africa); and Japan's trade surpluses with Europe and the United States were offset by energy and raw materials bills and capital exports to East Asia. Now, the global recession intensified interbloc rivalries relating both to the politics of surplus capacity – as in steel, petrochemicals and agriculture – and to the competition for leadership in the new technologies. Gunder Frank has characterized these tensions as follows.[19] To begin with, Western Europe and Japan increased their penetration of traditional US markets in Latin America, while the United States expanded its market share of traditional West European markets in Africa, and both the United States and Western Europe developed their attempts to gain openings in the traditional Japanese markets of East Asia, especially in and through China. There has also been considerable rivalry for markets between the three blocs in the Middle East – for Western Europe has a trade surplus with the Eastern bloc and the non-oil exporting Third World, but a deficit with the United States, Japan and OPEC. In addition, conflicts have arisen between Western Europe and the United States as the economic fruits of *détente* were gathered by the former and the latter sought to apply pressure on the Soviet Union, illustrated by the gas pipleline episode, the prohibition of 'strategic' exports and the manoeuvres on the debt rescheduling for the Eastern bloc.

In addition to this rivalry, the huge budget and trade deficits of the 1980s represented a unilateral unbalancing of these trade and capital flows on the part of Washington. In effect, the high interest rates in the United States constituted an unwanted 'Marshall Plan in reverse',[20] channelling European and Japanese funds to the United States. Meanwhile, the debt crisis reversed the direction of capital flows between the core and the (semi-)periphery: the commodity price collapse, combined with the high real interest rates, spread depression through much of the Third World. In turn, the depression in the periphery circumscribed the scope for an export-led recovery in Western Europe and reinforced its dependence on the US market. And as compared with recessionary Europe, the boom in the United States along with the dollar revaluation, as well as the interest rate differentials between the United States, Europe and Japan, meant that the US trade deficit continued to deteriorate. Finally, although Japan and the 'Four Tigers' (Taiwan, South Korea, Singapore and Hong Kong) were the principal foreign beneficiaries of the Reagan boom, this growth took place at the expense of a more regionally balanced regulation of demand among the East Asian economies.

The resulting Japanese trade surplus with the United States (and to a lesser extent that of West Germany) has been the source of considerable protectionist pressure on behalf of the mature sectors of the economy in Congress; in order both to staunch such demands and to attempt to rectify the trade position, the administration decided in early 1985 to devalue the dollar. This decision marked the opening of the second phase of Reaganism as identified by Parboni. From now on: '[The] administration [was] concerned that defence of mature sectors of the US economy – the ones most endangered by foreign competition – might trigger retaliatory measures by trading partners that affect American exports of advanced goods [and services where the United States] . . . has a strong interest in opening up world markets'.[21]

Under these circumstances, Reagan had considerable difficulty, as did Carter, in convincing the Japanese and the West Germans that reflationary policies to correct global trade imbalances would work. Thus there emerged powerful tendencies pushing towards the fragmentation of what had been an increasingly open trading order (world exports rose from 12 to 22 per cent of output between 1962 and 1984); for as each bloc found its internal problems easier to manage through a reduced dependence on international trade, pressure mounted for the effective replacement of the General Agreement on Tariffs and Trade by increased bilateralism.[22] Against this incipient regionalization of the world economy were (and are) ranged those considerable forces which benefit from free capital movements and unrestricted trade, the transnational corporations and transnational banks, to say nothing of the memories of the 1930s. Moreover, given the global pattern of the reorientation of US power sketched above, the long-run interests of the United States lie clearly in the maintenance of an open world economic order. This growing unilateralism of the Reagan years can be seen with special clarity in the case of US action in the oil industry.

Oil in the World Energy Market

The Future of World Energy Use

In the early 1980s the International Energy Agency constructed a series of projections concerning the future energy balances of the OECD bloc.[23] Within the area as a whole substantial import dependence for energy supply is expected to continue until 2000 and beyond; but within this overall picture there are important variations both by region and by fuel. Thus while IEA energy import dependence is expected to remain roughly stable at 20–25 per cent, North American (the United

TABLE 6.1   Energy self-sufficiency in IEA regions, 1973 and
1983: Proportion of domestic production in TPER (%)

	1973	1983
North America	87.9	90.1
Pacific	28.1	40.6
Western Europe	41.6	66.2

States and Canada) dependence will increase by about 30 per cent to
about 5–10 per cent; Pacific (Japan, Australia and New Zealand)
dependence will increase by a similar proportion but to 60 per cent;
and Western Europe's (including Turkey) dependence is projected to
remain roughly stable at 40 per cent. There will also continue to be
important regional variations in the fuel composition of total primary
energy requirements (TPER). While all regions rely on solid fuels
(primarily coal) for 20–25 per cent of TPER and all rely on nuclear
and other sources for about 11–12 per cent, North America has an oil
to gas ratio of 1.69; Europe, 2.96, and the Pacific region, 6.65. The
combination of these varying relative exposures to the international
market and the disparities in the commodity composition of total
requirement is reflected, in turn, in the projected import dependence
to 2000 by specific fuels: the Pacific will be extremely dependent for
oil (90 per cent) and highly dependent for gas (60 per cent), whereas
Western Europe will be highly (60 per cent) dependent for oil. In
addition, gas consumption is projected to expand faster than coal in
the Pacific region, whereas the converse is the case for North America
and Europe. And finally, in all areas the share of TPER provided by
nuclear energy is expected to double by 2000 (these were, of course,
pre-Chernobyl estimates), while allowing for the greater demand in the
Pacific the share taken by hydropower (and other alternatives) is
expected to increase by 50 per cent in all regions. Some of these
projections are summarized in tables 6.1 and 6.2.[24]

The figures and projections in the tables reflect the central strategic
initiatives of the IEA member states: namely, to diversify oil supplies
so as to reduce dependence on OPEC; to substitute other fuels for oil –
especially nuclear power, gas and coal; to conserve energy in industrial,
transport and household sectors; and to expand research into alternative
or renewable energy sources. However, energy forecasting is a notoriously risky business and some important qualifications must now be
entered. Thus Odell, for example, has offered a radically divergent

TABLE 6.2   OECD import dependence by fuel and region, 1983 and 2000: Net imports as a proportion of demand (%)

	Total		Oil		Coal		Gas	
	1983	2000	1983	2000	1983	2000	1983	2000
IEA (total)	21.6	23.2	55.9	48.2	-2.3	2.4	17.3	7.6
North America	5.4	9.9	34.0	26.2	-19.4	-5.3	5.0	3.6
Pacific	45.9	59.4	89.8	91.3	4.6	11.9	59.5	62.4
Western Europe	41.7	38.8	72.7	62.3	33.1	14.5	35.4	6.7

estimate of future trends.[25] From the mid-1970s energy use has expanded at about 2 per cent per year due to the recessions, conservation and changes in the sectoral composition of output. And after remaining relatively stable throughout the 1960s, during the decade 1973–83 TPER/GDP and oil/GDP ratios decreased by 2.1 and 3.8 per cent per year, respectively; and in 1984 while gross domestic product in the OECD bloc increased by 4.6 per cent, the above ratios continued to fall, albeit at a slower rate, by 0.5 and 1.6 per cent, respectively. Working with a projected OECD growth rate of 2.2 per cent over the period 1982–2000, Eden estimated that the TPER/GDP ratio would fall by a further 12 per cent.[26]

Overall, Odell estimated that, with conservative energy use policies and enhanced levels of energy utilization in the Third World, the global energy demand by 2000 would be between 12,000 and 14,000 million tonnes coal equivalent – that is, a 14–33 per cent increase on 1983. In other words, Odell's central case assumption (a 25 per cent increase) for global energy demand is roughly the same as that of the International Energy Agency the OECD bloc, and yet all authorities agree that, because of their respective positions on the energy/GDP curve, the rate of increase in the demand for energy will be lowest in the OECD bloc, higher in the Eastern bloc and greatest in the industrializing countries of the Third World. Under this alternative scenario, Odell argued against any further expansion of nuclear power.

In the case of oil, the West's non-OPEC oil production exceeded OPEC's exports by 1983, whereas in 1973 the latter had been almost twice as important as the former. Indeed, by 1983 OPEC's oil exports were little more than the West's gas production and less than its coal production. Drawing on his own research with Rosing and that of the

World Bank,[27] Odell suggested that OPEC's decline would continue and that its gas production would be constrained by lack of markets. Recognizing that the net expansion of the global coal industry has been modest, Odell drew attention to the rapid growth of open-cast, internationally traded coal and the stagnation or decline of deep-mined coal in Japan and Western Europe. And on the basis of James's estimates,[28] Odell argued that the international coal trade was likely steadily to expand while in supply and demand terms there was no reason why the real cost of coal should increase at least until 2000. With the further assumption that by the middle of the twenty-first century the supply of energy from renewable and benign energy sources would be feasible, Odell concluded against the expansion of 'relatively untested and still unproven nuclear power' and for fossil fuel sources as 'a relatively safe and lowest possible cost supply'.[29]

In addition to the different estimates of total energy demand, the IEA–OECD projections and those of Odell differ in two key respects: first, the IEA in parallel with most national authorities placed considerable emphasis upon the buildup of nuclear power, whereas Odell saw no such need; and second, the IEA reckoned on a rapid expansion of the international coal trade, whereas Odell estimated only a slow buildup. As can be readily discerned from the IEA's emphasis on nuclear power and the expansion of internationally traded coal, its preferred strategies rely very heavily upon both the extant strategies of the nuclear states and the predominantly US global energy and mining corporations that dominate these markets. In particular, Odell's strategy would require a greater degree of indigenous, state-sponsored energy production in the Third World than the free market dependence on the West's multinationals envisaged by the IEA and encouraged by the Reagan administrations working through the World Bank and other multilateral agencies.

Within the general global energy balance, oil remains the regulator of the world energy markets due to its versatility, convenience and even cost. None the less, to the extent that oil's market share was eroded by competition from other fuels and to the degree that OPEC lost market shares during the 1980s, OPEC came to be the marginal energy source by the mid-1980s. On this basis, Odell suggested that five regions could be identified which were likely to become 'wholly or largely oil-self-sufficient':[30] an East–Southeast Asian area, economically dominated by Japan but including the energy of Indonesia, Brunei, Thailand, Malaysia and Singapore; a region centred on the Caribbean which would also involve the participation of Venezuela, Ecuador, Mexico, Brazil and possibly the United States; an area of the Southern

Round, based on South Latin America, Africa and Australasia; the West European and Mediterranean basin, which has the lowest level of integration to date; and the North American region, comprised of the United States and Canada. As the regions gained in coherence, Odell predicted that the non-Middle East OPEC members – Algeria, Ecuador, Gabon, Indonesia, Libya, Nigeria and Venezuela – would face pressure to join their respective regional groupings. On the other hand, the Middle East OPEC states – Iran, Iraq, Kuwait, Qatar, the United Arab Emirates and Saudi Arabia – would either seek to prevent such groups forming or attempt to join that centred on Europe and the Mediterranean.

Although technical considerations do not rule out such developments, this general picture is unconvincing because it abstracts from the economic and strategic context within which the oil industry and market is situated. To begin with, Odell's East–Southeast Asian bloc is likely to include Australia and also China in the longer term: both have pressing needs to develop their relations with Japan; meanwhile the dependence of the latter on Middle East oil will continue far into the next century. This suggests, further, that given the capital-poor position of Africa, Latin America and Antarctica, the Southern Round will remain integrated into a global (if regionally skewed) market dominated by the transnational energy corporations. Moreover, the position of the United States in world energy markets, in conjunction with its refashioned and increasingly predatory hegemony, is most likely to underwrite a unilateral strategy buttressed by bilateral ties rather than regional cooperation. Finally, because of the continuing central role of the United States in the Middle East and notwithstanding Soviet energy exports to Western Europe, the basis for a European– Mediterranean region with OPEC participation is deeply fractured.

The World Oil Industry in the 1980s

In 1972 (the last full year before the major price increases), the Seven Sisters probably accounted for three-quarters of total oil industry profits.[31] By 1983 although the majors had lost control over the upstream end of the industry, they maintained a key role in marketing as an intermediary between producers and consumers. At this time, the five US majors and Standard Oil of Indiana were in the top ten American industrial corporations, while these five majors made one-seventh of all the profits of the Fortune 500 (roughly the same share as before the 1973–4 price rises). Moreover, in 1985 (the last full year before the price collapse of 1986) among non-US transnational corporations the five leading companies (by profits) were petroleum concerns – interestingly three of these were the national oil companies of Mexico, Brazil and Venezuela. In the sales league six out of the top ten were petroleum concerns – with a similar national oil company representation from Italy, Mexico and France. Also, in 1972 the majors controlled approximately three-fifths of the West's oil, a third of US reserves and some 90 per cent of Middle Eastern and Latin American crude production. But higher oil prices encouraged non-OPEC production and a spate of acquisitions, so that by the early 1980s production ouside OPEC overtook OPEC exports. Indeed, between 1973 and 1981 the North Sea, Alaska and Mexico increased production by 16 million barrels per day. Meanwhile, the majors increased their control over US reserves by a process of exploration and acquisition. The majors now had the majority of their oil reserves in non-OPEC states: Mobil about 50 per cent; Texaco and Gulf some 60 per cent; BP, Exxon and Shell about 65–70 per cent; and Socal over 80 per cent. The overall result was that at OPEC's lowest point in 1985, for example, the Middle East controlled 64 per cent of non-communist world's reserves but produced only 25 per cent of output, while Western Europe and North America held 7.5 per cent of reserves and accounted for 30 per cent of output. And by 1988 North American and West European production accounted for 24.6 per cent of the world total, whereas these areas counted only 6.7 per cent of reserves. While these policies undermined OPEC's market power in the short to medium term, this was at the expense of a rapid rundown of reserves to production ratios elsewhere. At the levels of production of the late 1980s, US reserves might last for a mere 10–12 years.

Developments in refining have seen a rise of capacity in Western Europe, a decline in the United States and an increase in the Middle East.[32] These changes were the result of the long lead times for refinery construction (5–6 years), which meant that capacity was still coming on stream in 1981, although it was estimated in Western Europe that no new capacity was needed until 1995. The move to light, high-sulphur crude hit the US refineries which were particularly suited to heavy, low-sulphur oils. And, with their oil rents, the Middle East states constructed a new, second generation of huge refineries. However, such changes should be viewed in the context of the overall postwar shift: prior to the Second World War Western Europe refined only one-third of its total oil consumption; by 1979 European refiners had excess capacity, while more than 85 per cent of OPEC's exports took the form of crude. Indeed, the majors remained dominant within the refinery industry as the much heralded move of the Middle East national oil companies into West European refining did not progress far. In 1986 this industry had an annual turnover of some $60 billion, after shrinking by nearly one-third in the previous decade. Through this huge restructuring the industry bcame more capital-intensive and flexible, so that the centre of power shifted once again to the majors.

The post-OPEC deintegration of the industry meant that refineries had to generate profit in their own right, a position rendered more difficult by the estimation that demand would grow more slowly in the future. Any excess capacity made the industry very competitive, and so refining demanded greater investments both to become efficient and to meet environmental standards. Further, as the recent (1986) price swings of crude have demonstrated, profitable refining required flexibility. The fall in prices was none the less in the majors' favour, as it has reduced the ability of OPEC to fund further expansion into refining – even Kuwait is no longer a serious threat, and most OPEC oil products are consumed domestically or exported to the Far East. Similarly, the independent refiners have found the competition too much. On balance, while the threat of overproduction is far from banished, the prospects for profitable refining look set to build on the favourable results of 1988–9. (There are, however, sceptics: James Bond of the World Bank's Energy Division has argued that refining can only be made securely profitable if it can share the rent of the more lucrative upstream sector. The refiners in turn have argued that this was true when there was considerable surplus capacity but that now that there isn't the industry's future is promising – refineries can differ in their efficiency and flexibility to the extent of some making $5 per barrel more than others.)

Opinion in the oil industry remains divided as to its long-term future and on the movement of prices in particular. This can be seen by contrasting the prognoses of the two largest majors, Exxon and Shell. The latter argued in the late 1980s that, because of a combination of low rates of growth of demand and continued high levels of investment in non-OPEC areas, prices will remain in the range of $10–20 per barrel throughout the 1990s. Exxon, by contrast, is banking on rising real prices during the 1990s and beyond. Thus for 1987–8 Exxon's net income was derived as follows: exploration and production, 72.4 per cent; refining and marketing, 9.4; chemicals, 14.4; Hong Kong power, 3.1; and coal and minerals, 0.7. In addition, the company claims that it can now produce synthetic fuels for around $30 per barrel from coal and oil shale (down from some $60 per barrel in 1982). In other words, Exxon remains committed to the traditional preoccupation of the

TABLE 6.3   US net petroleum imports related to consumption, by region (%)

	OPEC	Non-OPEC	Total
1973	17.3	17.5	34.8
1977	33.6	12.9	46.5
1983	12.1	16.2	28.3
1986	17.1	15.7	32.8

Source: Adapted from Norman S. Fieleke, The International Economy under Stress, Ballinger, Cambridge, Mass. 1988, p. 12, table 1.3.

majors with a hedge based on synthetic fuels. In fact, as we shall see, an emerging stability might increasingly characterize the industry, based on a marked commonality of interest between the United States, a few majors and the leading Gulf producers.

The Long-run Position of the United States in the Oil Market

The second oil shock raised prices in real terms – measured against industrial country exports – by 112 per cent between 1978 and 1981. The economic effects of this were broadly similar to those of the 1973–4 price increases, as was the action taken to conserve oil and diversify from OPEC sources. Thus between 1973 and 1987 OPEC's share of world crude production fell from 56 to 32 per cent while the share of oil in total primary energy consumption fell from 48 to 38 per cent over the period 1973–85. By the end of 1986, the IEA member states had acquired emergency stockpiles equivalent to four months' consumption. For 1973 to 1985, the oil/GDP ratio fell by 38 per cent in the OECD bloc as a whole, 38 per cent in West Germany, 48 per cent in Japan but only 29 per cent in the United States. An indication of the US position can be seen from table 6.3 which expresses its net petroleum imports as a percentage of total petroleum consumption.

Between the two oil shocks non-US OECD demand fell by 2.3 per cent while US demand increased by 12 per cent; and similarly, oil imports into the non-US OECD area (excluding Britain and Norway) fell by 2 per cent; yet US imports increased by around one-third. In consequence OPEC's share of US imports rose from one-half to over 70 per cent. Projections into the future depend somewhat on assumptions regarding the long-run trend of prices and the rate of discovery of new reserves outside of OPEC. But the industry consensus is that non-OPEC production outside the United States will at best improve marginally on 1985 levels by 2000 and at worst begin to decline, that US production will continue to fall, and that as a result the United States will become increasingly dependent on net imports.[33] In 1985 the US imported 4.2 million barrels per day; by 2000 this figure could be somewhere between 9 and 14 million barrels per day. Despite efforts both to expand production in Alaska in the hope of discovering another Prudhoe Bay (currently the source of about 20 per cent of US production) and to increase energy imports from Canada (now supplying some 5 per cent of US oil needs, 6 per cent of gas consumption and 2 per of electricity generation) through the United States–Canada free trade agreement of 1987, the United States might be importing 50–70 per cent of its oil by 2000. And given that non-OPEC production is expected to remain more or less stable, it is very likely that OPEC will come to dominate this increased level of imports. However, it is important to stress both that world oil trade is (largely) denominated in dollars, thereby mitigating any balance of payments constraint faced by the United States from this rising dependence, and that the projected US oil-import dependency remains low by comparison with either the European Community or Japan.

The United States in the Middle East

Introduction

Saudi Arabia is 'a stupendous source of strategic power, and one of the greatest material prizes in world history'.
                          State Department analysis of 1945[34]

The 'strategic and geopolitical significance of Saudi Arabia is quite likely second to no other nation on the face of the earth in its importance to the future well-being of the free world'
General P.X. Kelley, Commander of the Rapid Deployment Force, 1981

The umbilical cord of the industrialized free world runs through the Strait of Hormuz into the Arabian Gulf and the nations which surround it.
                          Caspar Weinberger, Secretary of Defence, 1981

After the declaration of the Eisenhower Doctrine, the state visit of King Saud to Washington in 1957 'moved [Saudi Arabia] away from the forefront of the "progressive" states opposed to western influence

in the Arab world and placed it firmly in the "conservative" flank. . . .
Saudi Arabia was now allied to the British-protected states of Iraq and
Jordan, and no longer to Egypt and Syria'.[35] A decade later, with
the defeat of the Egyptian, Syrian and Jordanian armies in the 1967
Arab–Israeli War and the announcement of impending British withdrawal from the region by 1971, Saudi Arabia's position in the Gulf
was strengthened still further. The twin pillar policy of 1969 cemented
relations with the United States. From this position, Saudi assistance
to the United States has taken a variety of forms: it provided finance
to aid Egypt's switch from the Soviet Union to the West; it financed
the airlift of Moroccan troops to Zaire, donated tens of millions of
dollars to North Yemen, and lubricated Somalia's turn to the West,
thereby securing a naval base for the United States, and more recently
it has supplied perhaps $2 billion to the Afghan rebels.[36]

However, as Stork has observed, after 1973–4 Arab politics consisted
of 'neither unity nor polarisation, but fragmentation and disintegration':[37] the political centre of the Arab world did not shift from Egypt
to Saudi Arabia; rather, it disappeared. None the less, US policy
continued to receive tentative backing from Tehran, Riyadh and Cairo.
Kupchan has explained this as follows:

> The formation of a moderate Arab coalition encouraged American diplo-
> matic efforts on the Arab–Israeli front, while simultaneously limiting
> Soviet influence in the region. The decline of Soviet leverage allowed
> the United States to devote more attention to regional issues, thereby
> strengthening the conservative coalition. At least temporarily, Israeli
> security, access to oil, and containment of the Soviets appeared to be
> complementary, not contradictory objectives.[38]

Rebuilding a Regional Coalition

By the late 1970s, however, a number of developments in the Middle
East and the surrounding regions posed a series of challenges to US
strategists.[39] Most generally, the overall political and strategic nexus
deteriorated. In Southern Europe as the dictatorships fell in Greece,
Portugal and Spain, Eurosocialism and -communism advanced (as it
did in Italy and France). In Ethiopia, situated in the strategically
significant Horn of Africa, the collapsing regime of Haile Selassie was
overthrown and replaced by a Marxist military regime. Also in 1978
the Khalq Party took over in Afghanistan. Most important of all, in
January 1979 the Shah fell in Iran, posing a dual challenge to US

policy: first, the United States had to make good the strategic role –
both regional and anti-Soviet – formerly played by Iran; and second,
the management of regional affairs was greatly complicated by the
instability attendant on the Revolution itself. The 'loss' of Iran was
particularly significant as it demonstrated the precarious nature of oil-
based and foreign-trasnsfer financed 'modernization'. Compared with
an equivalent revenue base financed by domestic taxation, oil revenues
facilitate the expansion of the clientelist and repressive apparatuses of
the state while the representative branches remain underdeveloped.
Iran thus provided a salutary lesson for Saudi Arabia (and even Iraq)
both as to its pattern of domestic development and in relation to the
political dangers of too close an alliance with the United States.

US leverage over Saudi Arabia was further weakened by the transformation of arms-as-aid to arms sales, and by the already high level
of sophistication of Saudi weaponry from past sales. An even greater
complicating factor was the ever-present Arab–Israeli conflict. After
the Iranian Revolution, Israel made clear that it was ready to assist the
United States in prepositioning and basing agreements, but with the
expulsion of Egypt from the Arab League after the Camp David
Agreement the Kingdom's room for manoeuvre in the Arab world
was closely circumscribed. Indeed, the combination of the Accords
(September 1978) and the Revolution meant that 'the Tehran–
Riyadh–Cairo alignment behind the pro-American consensus had crum-
bled'.[40] (For Israel the Iranian Revolution was both a strategic and an
economic blow, as Iran had supplied around one-half of Israel's oil.)
In response, the Carter administration sought to rebuild its position in
the Middle East. In February 1979 Secretary of Defence Brown publicly
stated that the 'U.S. [was] prepared to defend its vital interests with
whatever means are appropriate, including military force where necessary'. Indeed, planning for the Rapid Deployment Force (RDF) began
long before the Soviet invasion of Afghanistan (December 1979) or the
Carter Doctrine (January 1980), an illustration that the Force's task
was as much directed towards regionally generated conflicts as it was
to any putative Soviet agression directed at Iran. (The intraregional
concerns of US planners was a further factor militating against Israeli
co-operation.)[41] Still, events in Afghanistan did reduce the bureaucratic
resistance within the administration as well as providing the main
mission for the force – the defence of the Iranian oilfields from a Soviet
attack.

Now, despite the fact that neither Carter nor Reagan ruled out either
the use of nuclear weapons or the possibility of horizontal escalation
to defend oil, the principal focus of attention was an attempt to develop

a coherent conventional approach, the Zagros strategy. An essential precondition for the success of the RDF was US access to bases in the region; this required, in turn, a political rationale that made sense in regional terms.[42] Accordingly, in 1980 the Carter administration signed basing agreements with Oman, Kenya and Somalia, while in 1981 Saudi Arabia and Egypt agreed to co-operate on prepositioning and overbuilding. These gains were limited, and Haig's March 1981 attempt to orchestrate a regional anti-Soviet coalition was (in public at least) decisively rebuffed. Once again, the United States turned to military sales: specifically, in March 1981 Reagan recommended to Congress the sale of airborne warning and control system (AWACS) to Saudi Arabia; and after intensive lobbying and attempts to placate the pro-Israeli lobby the Senate narrowly (52 to 48) approved the deal in October. The significance of the AWACS deal was far-reaching:

> the AWACs sale circumvented one of the RDF's main obstacles: the need to have large quantities of American equipment and sophisticated weaponry in the region. . . . [The method] was to sell the equipment to the Saudis rather than preposition it on their territory. . . . [While for Saudi Arabia the attraction was the] potential both to enhance Saudi defense capabilites and to further attempt to establish a regional security network in the lower Gulf. . . . [And indeed] the Gulf Cooperation Council emerged simultaneously with the evolution of the AWACS strategy.[43]

The basing problem turned on the loss of surveillance capabilites directed towards the Soviet Union in northern Iran as well as the need to preposition large quantities of fuel to supply the aircover needed by the RDF. With its access to AWACS, and while Iraq and Iran were embroiled in conflict, Saudi Arabia was now in a position to constitute the centre of a regional alliance in the lower Gulf. And while the Gulf leaders could pledge themselves to keep the region 'free of international rivalry, especially in regard to the presence of naval fleets and foreign bases', the spilling over of the Gulf War into attacks on Kuwait (along with an Iranian-backed attempted coup in Bahrain and the defence pact between Libya, Ethiopia and South Yemen in August 1981), precipitated the formation of the Gulf Co-operation Council (GCC).[44] Against this background, then, the United States attempted to build a regional anti-Soviet coalition, strengthened its strategic ties with Israel, promoted open trade and investment and increased the direct US military presence. As Cypher shrewdly noted, the Reagan administration had a 'comparative advantage' over its allies in 'functional militarism': namely, the use of its military power to secure access to

Third World resources and markets.[45] Thus, the region witnessed an expansion of US bases and the creation of the Rapid Deployment Force (Central Command – Centcom – from 1983), while the conventional weapons buildup (especially the Navy) was in part directed at the Gulf.

From 1983 onwards Pakistan allowed US P-3 Orion aircraft surveillance planes access to bases so that they could survey the Gulf, the Indian Ocean and track the Soviet fleet, particularly its nuclear submarines.[46] Indeed, although Diego Garcia remained the central base for Centcom, Pakistan came to play an increasing role in US strategic thinking, becoming the third largest recipient of military aid after Israel and Egypt. In addition to its role in the supply operation to the Afghan guerrillas, Pakistan borders Iran and has a 400-mile coastline with the Arabian Sea. Already, after the 1971 war and the loss of Bangladesh, Bhutto had distanced Pakistan from the subcontinent and attempted to project it as a regional power in the Middle East. (Pakistan has military missions in over twenty states in Africa and the Middle East, making it the largest exporter of military manpower in the Third World.) In general, this involved cementing economic and military ties with the Gulf oil states by providing military expertise and personnel in return for economic assistance to rebuild the army. (US assistance helped to build up the navy.) More specifically, Pakistan has played an important role in the unification of the armed forces of the GCC members. The Gulf Co-operation Council itself, though nominally focused on external threats, was primarily a response to the common internal threats faced by the Gulf regimes and their growing economic integration. Indeed, as a result of the precarious and restricted social bases of these states, any challenge to the regime *is* a threat to the security of the state.[47] It is in this context that the Gulf War must be assessed.

The Gulf War

The 1975 Algiers Agreement between Iran and Iraq comprised three main elements: a resolution of a long-standing border dispute, an accord to divide the Shatt al-Arab waterway down the middle and a stance of non-interference in each other's internal affairs. Prior to this, Iran had promoted Kurdish revolt in Iraq with assistance from the CIA while Iraq supported elements of the Iranian opposition. (After his exile in 1964, Khomeini resided in the Shi'a city of Najaf in Iraq from 1965 until he moved to Paris in October 1978.) If Iraq's invasion of

Iran in September 1980 aimed to gain control of the Shatt al-Arab at a minimum and to overthrow the Islamic Republic and annexe the oil-rich province of Khuzistan at best, the overriding motive for aggression was political, not territorial, and lay in the political challenge which each state posed to the other.[48] After the initial Iraqi advance and the Iranian counterattack of 1982, the Gulf conflict settled into a brutal war of attrition in which Iran's 3:1 manpower advantage was balanced by Iraq's superior armour and airpower (and eventual resort to chemical weaponry). Lacking the domestic resources for a prolonged conflict, both powers, but particularly Iraq, sought to internationalize the war. And without massive arms sales by the superpowers and others – including France, Britain, Israel and China – neither side could have fought for so long. Iran's attempt to interdict Iraqi oil exports failed as Saudi Arabia and Kuwait sold oil on behalf of the latter and pipelines across Turkey and Saudi Arabia conveyed more oil for Iraq than Iran could sell. Further attempts by Iran to spread Islamic revolution in Saudi Arabia, attacks on Kuwaiti and Saudi shipping and support for the Kurds in Iraq also failed to weaken the Arab Gulf powers.

Although Saudi Arabia became Iraq's principal financial backer this was offset to a small degree by some supplies of refined petroleum products to Iran. In fact, for reasons of both internal (sizeable Shi'a populations) and external security, the Gulf states wanted to normalize their relations with Iran but could not allow its victory over Iraq. Equally, each superpower supported both the contending powers as the long-run position in Iran was clearly central to their competition in the region; and while Iraq was a Soviet ally, the United States would not comtemplate an Iranian victory with its consequent implications for the Gulf states. Thus, as soon as Iran had secured its borders in the summer of 1982, the Soviets began to support Iraq's calls for peace, and by mid-1983 both superpowers sought to preserve the Iran–Iraq balance by assisting Iraq. In addition, the United States renewed diplomatic relations with Iraq in 1984 and within three years the latter had become its third largest trading partner in the region, after Saudi Arabia and Egypt. Against this international stance, the balance of forces in Iran remained in favour of continuing the war; and in the conduct of the war, Iraq's military strategy was of central importance.[49] For it was above all Iraq that posed a threat to Western shipping in the Gulf, as its strategy involved an attempt to show Iran that the war was in fact unwinnable, by attacking oil processing and export installations as well as tankers in the Gulf together with a mounting assault on key points of the Iranian economy. However, notwithstanding the support of the Arab states for Iraq, in 1985 Saudi Arabia began cautious contacts with

Iran; by 1986 the United Arab Emirates and Bahrain (both with large Shi'a communities and trading links with Iran) could no longer ignore Iranian attitudes.

Meanwhile, 1985–6 also witnessed renewed Soviet diplomatic activity in the Middle East. The Soviet Union initiated diplomatic relations with Kuwait, Oman and the United Arab Emirates while sustaining those with its allies in Iraq, Syria, South Yemen and Libya. Furthermore, the Soviets opened semi-official relations with Israel, Jordan and Egypt, proposing an international conference on the Palestinian question. And finally, though continuing to supply arms to Iraq, the Soviet Union commenced economic discussions with Tehran, signing a treaty of friendship and co-operation in October 1987. The 'Irangate' scandal (November 1986) only served to underline the absence of a coherent US policy in the region. The growing insecurity of Kuwait in particular, however, was to provide the pretext for a temporary substitute. For even if Kuwait and Saudi Arabia pressured the United States into reflagging Kuwaiti tankers by threatening to sell Treasury bonds,[50] the basic impetus behind the operation was to keep the Soviets out of the Gulf. This framework also served a range of more concrete goals. In part, the reflagging mission, crafted to secure bipartisan support, was a further step towards securing a domestic consensus around increased intervention in the Third World. The fact that the United States gained a mere 7–9 per cent of its oil from the Gulf in 1980, as compared with figures of 75 and 50 per cent for Japan and Western Europe respectively, is not of direct relevance – though some estimates suggested that by the mid-1990s OPEC oil would count for 50 per cent of US energy imports. Rather, understood as the management of hegemony, US action in the Gulf was guaranteeing the economic stability of the West as a whole through that structure in which it remained the preponderant power, the world military order. And lastly, by the protection offered to Kuwaiti and Saudi shipping, the United States hoped both to establish a permanent military presence in the region and to pressure the Arab states for greater access to bases.

By this state of the conflict, the costs imposed on the populations and economies of the combatants was immense. The economic costs of the war to early 1988 can be seen in table 6.4.

To make matters worse, through 1987–8 Saudi–Iranian relations deteriorated, with the riots at the *hajj* in July 1987 prompting an open breach; and continued Iranian attacks on Saudi shipping alongside Saudi suspicions that Iran was promoting the spate of bombings in the Kingdom resulted in Riyadh ending diplomatic relations in 1988. At

TABLE 6.4   The economic costs of the Gulf War (US$ billion)

	Iran	Iraq
Revenue/expenditure losses		
Loss of oil revenue	23.4*	65.5*
Extra military expenditure	24.3	33.0
GDP losses		
GDP loss (oil sector)	108.2	120.8
GDP loss (non-oil sector)	30.3	64.0
Fixed capital losses		
Unrealized capital formation	76.5*	43.4*
Loss from destruction (oil sector)	n.a.	n.a.
Loss from destruction (non-oil sector)	25.9	8.2
Total	188.7	226.0

* Included in GDP losses, therefore do not add when summing totals.
Source: See Andrew Gowers, 'The balance tilts against Iran', in the *Financial Times*, 27 April 1988.

the same time, 'the Mecca crisis was a catalyst in the process of unifying divergent Arab positions',[51] allowing the Arab League both to offer clear support for Saudi Arabia and Kuwait and to renew relations with Egypt. By this point, the combination of external (the United Nations Security Council had unamimously passed Resolution 598 calling for a ceasefire in July 1987) and internal pressures on Iran at last brought the conflict to a halt in July 1988: from Iran's standpoint 'it was necessary to end the war in order to save the Islamic revolution'.[52] Thus, the responses of the Arab states and the United States to the endogenous political and military conflict in the Gulf mediated an overall increase in US influence in the region. On the one hand, the distraction of Iraq by the War, combined with the threats to the Gulf states, produced the GCC. And on the other, the US intervention in the conflict to forestall any Soviet diplomatic advantage, combined with the US sale of AWACS to the central power of the southern Gulf, Saudi Arabia, increased US power in the region. In turn, this growing alignment – although it publicly took an independent stance *vis-à-vis* the superpowers – underpinned the rising dominance of the Gulf producers within OPEC.

OPEC: 1979–1989

After the instability of 1979–80, 1981 proved to be a year of price reunification for OPEC. By early 1982, however, Iran had cut its price in an attempt to expand its market share and so increase revenues for war-financing. Meantime, in conditions of depressed demand the effects of Saudi attempts to discipline the price-hawks had resulted in a glutted market. As a result, at the sixty-third conference (March 1982), 'Opec . . . finally agreed to turn itself into a cartel'.[53] Iran and Iraq were both given quotas of 1.2 million barrels per day; and although Iran refused to accept this, the depredations of the war meant that neither could in practice exceed its quota. (Indeed, the Gulf War was in this respect fortunate for OPEC: before the conflict Iran produced 5.5 and Iraq 3.5 million barrels per day, and so the war 'involuntarily contributed 6.5 mb/d to the total OPEC reduction of around 12.5 mb/d between 1978–82'.)[54] A number of uncertainties and strains soon eroded this agreement. The spring of 1982 witnessed a renewed Iranian offensive in the Gulf War and Syria closed Iraq's pipeline, thereby cutting off 400,000 barrels per day, while Iran's production continued to increase to 2.8 million barrels daily in December. On the political front, in June Israel invaded Lebanon and King Khalid died and was succeeded by Crown Prince and Prime Minister Fahd. Of even greater importance than these essentially contingent worries was the fact that OPEC was attempting to determine simultaneously prices and output: in addition to setting an overall production ceiling of 18 million barrels per day, the sixty-third conference confirmed $34 per barrel as the marker price.

At this point, the interaction between OPEC and the growing North Sea (NS) production became apparent through the pressure placed on Nigeria.[55] For as NS output expanded outside the OPEC pricing structure, so it conquered US and West European markets. With the Gulf crudes losing their market share and yet with Nigeria's prices tied to those of OPEC, Nigeria became the swing producer in the Atlantic Basin. In the absence of the cartel arrangement, therefore, Nigeria might have been compelled to leave OPEC, as indeed the relevant oil companies may have been urging it to do. In this context, in January 1983 Yamani predicted that NS prices would be cut, that this would place pressure on Nigeria again, and that OPEC would then have to reduce the price of its marker crude. February witnessed price cuts by the British state oil trading company, BNOC, and by March (sixty-seventh conference) OPEC had a second agreement, with Saudi Arabia now playing the role of the swing producer. This pattern continued throughout the mid-1980s. In October 1984 the Norwegian state oil company, Statoil, tied its prices to those of the spot market and BNOC followed, presaging the breakup and privatization of the latter in 1985. Once again Nigeria was forced to cut prices. By now OPEC quotas were some 50 per cent of the volumes produced in 1977, and given the erosion of price increases by inflation the real incomes of the member states had actually fallen. In order to maintain a degree of unity Saudi Arabia reduced production to merely 2.4 million barrels per day by May 1985.

These attempts by OPEC to regulate prices by responding to demand changes through quantity adjustments over the period 1979–85 became increasingly difficult as non-OPEC supplies rose and world demand stagnated. (In 1986 world demand was still 10 per cent below the level of 1979.) During the first half of 1985, as its production fell to one-half of its quota and perhaps one-quarter of its potential (a mere 2 million barrels per day) Saudi Arabia was exporting little oil to the United States, was only a middling supplier to Europe and had fallen to second place in the Japanese market. At this juncture Saudi Arabia, Kuwait and the United Arab Emirates favoured a price of around $18 per barrel in order both to increase sales and to forestall further oil exploration and production in high-cost, non-OPEC areas, while such states as Iran, Iraq and Libya with lower reserves to production ratios wanted to maintain prices of around $30 per barrel. Accordingly, when in July 1985 Mexico began to cut prices, the Kingdom followed suit in September. Faced with Iraqi demands for larger quotas (and parallel claims on behalf of Iran) as well as widespread cheating, the Saudis adopted netback contracts in order to drive down prices and increase their market share. This pushed OPEC output up to 20.3 from 16 million barrels per day; in response prices fell from around $30 to $10–12 per barrel. Specifically, in July 1986 Saudi production stood at 6 million barrels per day and the spot price of Brent crude had fallen to $8.75 per barrel. The Kingdom thereby regained its position as the second supplier to the United States (after Mexico) and played a major role in supplying the international oil companies. The overall position in 1978 and 1985, which in turn prompted the Saudi actions, can be seen in table 6.5

The motivation behind Saudi Arabia's policy appears to have been both financial and organizational: on the one hand as the swing producer the Kingdom was paying a very high price, while on the other it was in danger of losing its directive role within OPEC. More generally, the financial position of even such low absorbers as Saudi Arabia, Kuwait and the United Arab Emirates – who together accounted for

TABLE 6.5   Oil production and consumption, 1978 and 1985

	Millions of barrels per day		
	1978	1985	Change (%)
Non-communist world consumption	50.3	45.3	-10
Non-OPEC production	18.6	25.3	+36
OPEC production	29.8	15.4	-49
Saudi Arabia	8.3	3.2	-61
Iran	5.2	2.2	-58
Iraq	2.6	1.4	-46

Source: Ian Skeet, OPEC: Twenty-five years of prices and politics, Cambridge University Press, Cambridge 1988. Adapted from pp. 211–12, tables 11.1 and 11.2.

over 80 per cent of OPEC surpluses – had deteriorated markedly by the mid-1980s.[56] In 1981, for example, the combined earnings from the oil exports of Saudi Arabia, Kuwait and the United Arab Emirates was $150 billion but in 1985 it was only $45 billion. By 1986 Saudi Arabia was expecting its fourth current account deficit in succession, and though imports were cut and development projects curtailed, its overseas assets were steadily being drawn down. Had prices remained at the level of spring 1986 throughout the year, then OPEC's annual income would have fallen by two-thirds. Accordingly, in August an agreement was reached limiting total output to 16 million barrels per day and on Iran's initiative Iraq was allowed to remain outside the system. (Indeed, by 1986 Iraq had nearly exhausted $35 billion of reserves, received an estimated $30 billion from Saudi Arabia and Kuwait and continued to benefit from their sale of 310,000 barrels per day on its behalf.) In October Yamani was sacked from his post, signalling a significant change of policy by the Kingdom: once again there was to be a target price ($18 per barrel) and a production ceiling (16.6 million barrels per day).

Iranian co-operation did not indicate a genuine political accord with Saudi Arabia but rather its determination to increase revenue by raising the price of its relatively fixed output. More significant was the backing given by the US administration. With US domestic production costs ranging from $3 to 15 per barrel, the low prices of 1986 resulted in severe difficulties for the southern oil states. In addition, looking to the longer term given the small reserves to production ratio of the lower forty-eight states, US policy has tried to encourage stability within OPEC at around $18 per barrel. However, by the summer of 1988 OPEC was once again in predictable disarray. The Saudi strategy had been premissed on the belief that a lower price would increase demand and that, in turn, this would buttress the quota system. Demand did recover, but not enough to offset the widespread cheating by Kuwait and the United Arab Emirates and the refusal of Iraq to accept its quota. Saudi Arabia now increased production above its OPEC quota of 4.3 million barrels per day to a reported 5.7 million barrels daily in October 1988.[57]

Above all, the ending of the Gulf War posed a fundamental, if medium-term, obstacle to the internal cohesion of OPEC. Other things being equal, Saudi Arabia's influence within OPEC depends on its room for manoeuvre *vis-à-vis* output; but with Iraqi and Iranian production increasing rapidly this asset will provide significantly less leverage. Thus large-scale purchases of oil-loading equipment, together with increased pipeline capacity to the Red Sea and the reconstruction of war-damaged terminals in the Gulf, mean that Iraq's export potential will rise from some 3.35 million barrels per day in 1989 to 5.5 in 1990. (Its 1989 OPEC quota was 2.78 million barrels per day for the fourth quarter.) Equally, in time there is no reason why Iran's capacity (3.5 million barrels per day in 1989) cannot match its former output of some 5 million barrels per day. Now, although Iran and Iraq have huge foreign debts to pay off and large infrastructural commitments to finance, increasing production could well be short-sighted as prices would doubtless fall. But substantial production potential on behalf of Iraq and Iran enables these powers to challenge Saudi 'moderation' over prices, and this may explain why the Kingdom sometimes allowed prices to rise above $18 per barrel in early 1989. Yet with a production ceiling of 19.5 rising to 20.5 million barrels per day in the fourth quarter, together with excess production – especially by the United Arab Emirates and Kuwait – totalling 2–3 million barrels daily, Saudi Arabia's refusal to reduce its quota below one-quarter of the OPEC total, and the absence of an agreement for 1990, OPEC was hard pressed to meet its target price of $18 per barrel.

On the other hand, Saudi Arabia initiated another change of strategy in June 1988 when it bought a half-share in Texaco's network in the Eastern United States for $800 million. And in November the Oil Minister, Hisham Nazer, announced that the Kingdom intended to link its production decisions to refining and marketing opportunites in the consuming nations. Kuwait, Venezuela and Libya have already achieved some success with this strategy, and the continued overproduction and resultant price instability appears to have persuaded the Saudis to follow suit. The consequence of this is that increasing volumes of crude will enter refineries without any reference to the OPEC target price. Combined with Saudi Arabia's refusal to act as the swing producer, these developments could signal a renewed integration of the majors and (some of) the producer states which would permanently banish the spectre of radical price swings in either direction. In reality, of course, although OPEC is often characterized as a dominant-firm cartel in which Saudi Arabia (perhaps with its allies Kuwait and the United Arab Emirates) plays the role of the swing producer, the Kingdom never achieved this degree of dominance within the organization. Rather, between 1982 and 1985 OPEC was best understood as a partial market-sharing cartel, and with a simultaneous policy of price-fixing a weaker arrangement would be hard to specify. In these circumstances, the recent changes in Saudi strategy might signal a shift in the balance accorded to each arm of its hybrid role, towards increasing the integration of the Arab Gulf producers (perhaps also involving a *modus vivendi* with Iran) into the West at the expense of a clear directive role within OPEC as a whole. And given the fact that by the mid-late 1990s only Saudi Arabia, Iraq, Iran, Kuwait and the United Arab Emirates will remain major OPEC exporters, such a strategy has a clear rationale.

Traditionally, the Saudis have sought to strike a balance between disciplining OPEC production levels in favour of price stability and staying the power of the organization's more assertive members. Two developments have rendered this balancing act increasingly difficult for the moment: first, the end of the Gulf War has dramatically expanded the production potential of OPEC as a whole, thereby weakening Saudi leverage; and second, the emergence and expansion of a significant trade in natural gas, combined with the absence of any OPEC procedures in this regard, demands nationally based solutions. Linking petroleum production to marketing and refining opportunities both in the West and in Saudi Arabia's own development projects might provide a stable alternative, one which will further link the Kingdom's interests to those of the leading Western economies (see below). For in reality while the United States and the majors have often signalled a sharp antipathy to OPEC, this is strictly for public consumption in the West. And if both would be happy to see the non-Gulf members gradually detached from the organization, over the medium to long term a stable, non-market arrangement between the West and the Gulf producers could provide a renewed mechanism to avoid periodic crises of overproduction and thus replicate the stability of the period 1928–1973. Such a sub-imperialism in the Gulf would in turn be premissed on a direct linkage of economic development in the relevant states to circuits of capital dominated by the transnational companies of the West, thereby precluding a more autocentric pattern of development in the Arab world as a whole. A factor that has (potentially) run against this tendential integration, however, has been the attempt to diversify away from OPEC oil.

The North Sea in the World Oil Market

For a number of reasons, North Sea (NS) petroleum production played a larger role in the world oil market during the 1980s than its share of reserves would suggest.[58] This production rose rapidly during a period of stagnant world demand and, in tandem with Mexico and Alaska, partly displaced Middle East OPEC supplies. As a result, the North Sea occupies a key role in the international energy market: its crudes have pentrated the Northwest European and the US markets while its spot and forward prices play something of a barometric role. (Paper trading of Brent crude is some ten to fifteen times physical transactions, due to 'tax spinning' – arm's length sales for fiscal purposes.) But in so far as OPEC can react to changes in oil demand through quantity adjustments, the NS market is not a significant actor of medium-run price formation. The contemporary oil market in the short to medium term is two-tiered: outside OPEC changes in oil demand result in price adjustments but for most of the larger OPEC producers quantity adjustments are the norm.

The four largest operators in the North Sea are Shell (in 50/50 partnership with Esso in all UK fields), BP, Mobil and Phillips, together accounting for 65 per cent of all acreage. Similarly, the four largest producers – BP, Shell, Esso, and Statoil – account for nearly half of the total production, while the top ten account for over two-thirds. On the supply side of the market, BNOC, BP, Shell, Esso and Statoil dispose of nearly 70 per cent of total output. (The Herfindahl index – the number of equal-sized companies that would produce the same degree of market concentration as the actual distribution of companies with their unequal shares – on the supply side was 5–6; this has probably increased to 10–11 with the abolition of BNOC.) On the demand side, the largest six buyers account for 64 per cent of output (a Herfindahl index of 8, 9 after the abolition of BNOC). In other words the abolition of BNOC increased the oligopoly power of the buyers relative to the sellers; or, since the six largest buyers are Esso, Shell, BP, Texaco, Mobil and CFP, the power of the majors with respect to both the independents and the (privatized) national oil companies.

The principal export markets for NS crude during the 1980s have been Germany, Holland, France and the United States, which together accounted for some three-quarters of output. Indeed, by 1985 and largely as a result of the fall in OPEC production, the North Sea was the third largest producer (after the Soviet Union and the United States) and the second largest exporter (after Saudi Arabia). Thus, as Mabro et al. have concluded: 'The centrality of the North Sea market, resulting from the size . . . from the price adjustment characteristics of its mode of operation, from price transparency and from the centrality of the crudes themselves in the crude oil supply-mix, enables it to play a barometric role'.[59]

If the political security of the North Sea is also considered, then its role in the IEA strategy of diversification away from OPEC is obvious. Furthermore, due to this strategic importance of the North Sea, both the majors and the IEA states – especially the United States and Britain – have a continued interest in higher than necessary oil prices (that is, those prices that could meet world demand if the lowest cost sources were fully operational). In the early 1980s a series of changes in the NS tax regime, notably in the British sector, resulted in the *Financial Times* reporting in 1985 that these alterations 'make the North Sea one of the most attractive areas in the world for hydrocarbon exploration and development'. And although the price collapse of 1986 led to a considerable cutback in NS activity, substantial technical progress and cost-cutting rendered any price above $12 per barrel viable by 1988.

In fact, throughout the world the offshore industry had made a significant recovery by the late 1980s. Even with prognoses of prices as low as $15 per barrel in real terms for the 1990s, vigorous investment programmes were forthcoming as was a strong market for oil assets in such mature offshore sectors as the Gulf of Mexico and the North Sea. Exploration costs have been dramatically reduced by technological progress in geology and computing which now enable oil to be located in smaller pockets with a high degree of precision. Similar if less spectacular advances have been made with respect to production costs. By these means, the attempt by OPEC to thwart the expansion of non-OPEC production has been defeated since at prices below $15 per barrel the financial pressure on OPEC will dictate production cuts. Over the longer term, non-OPEC supplies are probably at best a stopgap, for most estimates suggest that sometime in the 1990s the world oil market will shift in favour of the Gulf producers. This is another factor underpinning the alterations in Saudi strategy noted above.

OPEC, the Arab Economy and the World Economy

> Arab history does not, of course, begin with oil, nor does the history of
> its integration as a periphery dominated by imperialism. None the less,
> oil has certainly accentuated the distortions of dependent Arab develop-
> ment, both before, and especially after, 1973. . . . . The apparent wealth
> provided by oil from 1973 onwards, has of course, accelerated growth,
> but in a regressive direction; it is indubitably the root cause of the Arab
> world's increasingly unequal integration into the world system.
>
> Samir Amin, *The Arab Economy Today*[60]

The shift in the balance of political and strategic alignments in the
Arab world consequent upon the Arab defeat in the Six Day War was
underwritten by 'the ascendancy of the oil-producing states in the
region and, through them, the region's closer integration into the
international capitalist system'.[61] The dominance of oil capitals alongside
the related banking interests has engendered a distinctive economic
structure based on an integration of oil-producing and labour-exporting
economies.[62] While domestic capital investment has been directed principally at infrastructure, Gulf industrial activity has focused on export-
oriented industries in which there was already substantial global surplus
capacity (above all, petrochemicals). In consequence, intraregional
trade comprises a mere 10 per cent of the total (the comparable figure
for Western Europe is some 50 per cent) and less than 5 per cent of
the foreign investment of OPEC states is intraregional. As a result of
this low level of regional integration and externally oriented development, 'the Arab world today is neither a political nor an economic
unit. Each of the Arab states is integrated as a separate unit into the
world captialist system'.[63]

Moreover, in addition to a relative absence of forward and backward
linkages in industry, the Arab economies suffer from a pronounced
backwardness in agricultural production. Unbalanced growth has also
resulted in a severe misallocation of labour power within the Arab
economies, as has the resistance to female employment in many states.
In Saudi Arabia, for example, the rapid pace of growth demanded a
huge influx of foreign labour, skilled from the West and unskilled with
racist pay scales from the Third World. (The Saudi population is some
6 million – of which some 60 per cent are under 21 years of age – and
an additional 2–2.5 million expatriates work in the Kingdom.) During
the prolonged recession of the late 1980s, the Kingdom was thus able
to protect Saudi personnel by drawing down on foreign reserves and
cutting the benefits and pay of the immigrant labour; but there is a

growing tension between the continued need for expatriate labour and Saudi unemployment (especially among women). Finally, at a macroeconomic level, the distribution of income in the Middle East has grown more unequal with oil-backed wealth: in 1970 the top decile consumed 37 per cent of national income; in 1979, 57 per cent; and in 1981, 62 per cent.

A principal aspect of this overall pattern of development can be seen again in the important case of Saudi Arabia, in which an industrial policy with two principal elements has guided the public sector investment decisions that dominate the economy.[64] Through a combination of the Royal Commission for Jubail and Yanbu and a management contract with the Bechtel Corporation, two new industrial cities are being developed, involving oil refining, petrochemicals, steel production and related service industries. And Petromin is developing a gas-gathering and treatment project to provide energy for the cities and inputs for the industries. By offering an assured supply of oil in return for investment (500 barrels per day per $1 million), the largely state-owned Saudi Basic Industries Corporation (Sabic) has signed a number of joint ventures – including deals with Dow Chemicals, Mitsubishi, Shell, Exxon, Mobil and Taiwanese Fertilisers – with transnational corporations.[65] On this basis, it has been estimated that Saudi petrochemical production will amount to 4 per cent of world production by 1990 or 10 per cent of Western Europe's demand. Thus far Saudi joint ventures outside the Kingdom have been limited but they are likely to spread, as costs of constructing petrochemical plants are considerably cheaper in the West and deals in the United States, Europe and Japan would also overcome protectionist threats.

In the longer term, the Kingdom has plans to increase its oil production potential to accommodate the expected tightening of the market in the 1990s. Thus in 1989, Saudi Aramco began preparations for a $15 billion investment programme to increase crude output capacity to 10 million barrels per day by the mid-nineties. This will further consolidate the new form of linkage to the West in that the leading consuming nations are being encouraged to finance the planned expansion of Gulf OPEC oil production, guarantee transnational corporation investments and cease taxing away the benefits of lower oil prices.

Meantime, the level of arms exports to the Middle East continued to rise until the mid-1980s, thereafter falling somewhat. Within the total, the United States steadily lost market shares to the West Europeans, especially Britain, as Congress repeatedly blocked sales to Arab states and arms sales displaced direct, superpower-dominated aid. The general picture can be seen in table 6.6.

TABLE 6.6   Arms exports to the Middle East

Period	Total ($ million)	Percentage from		
		The US	The USSR	W. Europe and China
1964–73	9,447	34	50	6
1974–78	29,000	48	26	19
1979–83	63,355	22	31	30

* See Robin Rowley, Shimshon Bichler and Jonathan Nitzan, 'The Arma-dollar–Petrodollar Coalition and the Middle East', Working Paper 10/89, Department of Economics, McGill University, Montreal, p. 13, table 2.

More generally, the combination of an increasing number of suppliers in the world market, a rise in research and development and production costs and the overall contraction of demand created a space for increased leverage on behalf of the arms purchasers. On this basis, the buyers have attempted to create an indigenous defence and industrial base through extracting significant technological spoils from the sellers. In the case of Saudi Arabia, its military expansion has continued apace – albeit at a lower level in the late eighties. Military contracts – the Peace Shield programme with the United States and the Al Yamama deal with Britain – have been directly tied to compensatory agreements and barter trade. For example, the industrial offset programme man-aged by Boeing Industrial Technology Group includes a range of pro-jects from aerospace, through electronics, computing, telecommuni-cations and power technology to biotechnology and medical products. Such an arrangement bolsters the arms exports of the seller while 'the industrial offset programme is structured to facilitate import substi-tution, "high technology" transfer, export expansion and diversifi-cation, and employment opportunities'.[66] The barter component of these deals – pronounced in the Al Yamama project – in part pays for the armaments by crude oil liftings, thus further structuring oil transactions outside OPEC arrangements. (British liftings will amount to some 400,000 barrels per day, perhaps more if prices remain low and Saudi military demand high.)

US Dominance over World Oil Reconsolidated

Some of the salient contrasts between the differing directive roles of the United States in the postwar global oil order can be seen in table 6.7.

As should be clear, then, the restricted optic of realism – focusing primarily on the nationally based power indicators of states – renders invisible the structural power which derives from advantageous positions within the various dimensions of the international system.[67] OPEC did not break the overall directive role of the United States in the global oil industry any more than the postwar recovery of Western Europe and Japan, together with the end of Bretton Woods, undermined US economic leadership. Equally, the moment of OPEC no more betokened the loss of a directive role by US capital and the US state in the industry than the price rises of 1973–4 caused the recessions of the 1970s. What certainly had changed was the form taken by the global power of the United States; this altered markedly: in short, institutional preponderance has diminished while structural dominance has been extended. But to focus only on the loss of national relative advantage, while neglecting the focal structural role of the United States, leads to a series of symptomatic misreadings in which the decline of US power is routinely exaggerated. Specifically, while the share of the non-communist world's petroleum reserves held by OPEC is over ten times that of the largest eight international oil companies, these firms have extended their technological and financial supremacy since the mid-1970s. And so, on the basis of any likely medium-term movement in prices, as the Gulf states in particular need to upgrade their oil installations, discover new finds and even seek development capital, the majors are strongly placed to meet this need. Thus, as the international companies exhaust their reserves, a new symbiosis is set to emerge.

More generally, the core basis of US structural dominance lies in a number of features whose importance can clearly be seen in the domain of oil. In terms of the US position in the world economy four features stand out. First, there is the general global spread of US transnational corporations across the world market, evidenced in the fact that by the late seventies some nine-tenths of US overseas activity took the form of foreign production (compared with a figure of one-tenth for Japan). And this overseas capital investment accounts for nearly one-half of the total stock of accumulated foreign investment. In the Middle East and the oil industry, the position of US companies in the industrial

TABLE 6.7   The changing directive role of the US in global oil industry

Postwar period	*Reconsolidation of 1980s*
Economic position of the Middle East	
Coordinated, interterritorial control by the majors of Middle East oil production, fed directly into cartel-like, refining networks in the consumer nations	Nationally based, producer-state (NOC) determination of Middle East oil production: at first, fix-price contracts with anarchic output levels; increasingly, flex-price, netback contracts with Western consumers, involving joint ventures with the majors – at least among dominant Gulf producers
Limited economic development among Middle East oil producers; ruling families dependent on oil revenues	Growing, oil-financed industrial development; ruling families become ruling classes; state dependence on oil wealth; in addition Gulf financial assets predominantly dollar-denominated in Western financial institutions; and Gulf economies seek high-rent technologies from the West (joint ventures)
Political and military position of the Middle East	
Middle East states depend on direct arms transfers from the superpowers; some formal pro-Western alliances, some treaties with the Soviet Union	Middle East states (especially oil-rich Gulf states) depend on arms purchases from Western producers; Gulf Co-operation Council, Saudi Arabia and AWACS provide regional security
Gulf states emerging from direct political control by Britain; followed by short-lived twin-pillar policy centred on Iran and Saudi Arabia	Differentiation among OPEC states, rising dominance of Gulf members whose military security guaranteed by the United States
Radical nationalist regimes pose limited but real challenge to Western dominance, some Soviet support for radical regimes	State-capitalist regimes dominant; limited Soviet linkages to oil states; no reliable Soviet allies with significant oil production
United States in global oil market	
United States largely self-sufficient in oil; Western Europe and Japan dependent on US-sourced Middle East oil	Growing US oil import-dependence, but still low compared to Japan and the EC, and oil trade denominated in dollars

and military development of Saudi Arabia as well as the role of US capital in Western refining networks remains far ahead of that from the European Community and Japan (though late 1989 saw the opening of talks between the Community and the Gulf Co-operation Council for a free-trade agreement). Second, there is the continuing dominant role of the dollar in international financial and monetary relations, with nearly two-thirds of world reserves held in dollars in the late 1980s. The financial surpluses of the states of theGulf are primarily placed in dollar-based assets. Third, there is the still central position of the dollar in international trade, exempting the United States from the balance of payments constraints faced by other economies. This means that rising US oil imports do not pose the same challenge as does the import dependence faced by Japan and the European Community. Fourth, there is the absolute size of the US economy within the world market, resulting in both the low degree of foreign trade in relation to gross domestic product and in the relative raw material self-sufficiency of the North American bloc. Specifically, the energy (and oil) import dependence of the United States, although rising, is low as compared with that of Western Europe and the Pacific.

Within the world military order the US position also remains highly privileged despite a declining relative share of global military spending. First, despite the Soviet advance to rough strategic parity, the power projection capabilities of the United States outdistance those of the Soviet Union, and the technological superiority of its weaponry gives the United States a clear advantage in the international arms trade.[68] The creation of the Rapid Deployment Force and the AWACS deal with Saudi Arabia are instances of these capabilites. Second, the alliance systems of the United States and its bilateral relationships with regional sub-imperialisms extend far beyond any advantages the Soviet Union might gain from either its allies or Cuban and East German 'proxies'. In the case of the Middle East, for example, compare the US–Saudi Arabian relationship with that between the Soviet Union and Syria discussed in chapter 5. Third, within the Western camp itself, although European arms sales have expanded considerably (especially in the Middle East), there is as yet no sign of the European Community displacing the regional role of the United States. Indeed, the events in the Gulf during the Iran–Iraq War indicate precisely the converse: namely, that the West European Union will tie Western Europe's military capabilities closely to US leadership in Third World deployments.

Geopolitically speaking, the reorientation of US economic and military power in relation to the global oil industry and the Middle East has not taken the form of an attempt to reimpose direct control over the region's resources and to create a forum equivalent to the Cento of the 1950s. Rather, the position of the United States has been reconsolidated through the attraction of radical nationalist, but state capitalist, regimes to the world market, together with the bilateral economic and military ties noted above. Combined with the US role in the Arab–Israeli dispute, these pressures were also reflected in the failure of the Soviet Union to construct a radical regional coalition. Thus, the collapse of Arab socialism and the relative eviction of the Soviet Union from the Middle East have given the United States a significant role in regional politics. Furthermore, following the partial eclipse of French and British imperialism in the Middle East, the growing scale of EC trade and arms sales to the region have not resulted in a corresponding increase in European political influence.

Conclusions

Throughout this study I have argued that the vantage points provided by realism and orthodox economics are singularly ill-suited to apprehend the contours of global politics. Instead, I have developed a conjunctural model of world politics drawing both on Marxian accounts of the capitalist structure of the modern world economy and on recent theoretical work on the administrative power of the modern state and the associated structure of sovereignty in the community of nation-states. The relation of world oil to US hegemony has served as something of a test case, in so far as it has been this domain which, by common agreement, has witnessed the most precipitous of changes. In this specific case, the *strategic* character of oil as a commodity in the epoch of postwar hegemony is rendered invisible either by a realist concern with the national resources of separate powers, or by the neoclassical inability to theorize economic power in the world economy. By contrast, a conjunctural analysis, placing the development of US oil policy in the context of its wider management of the world economy and nation-state system, demonstrates the central role which oil played in the organisation of US hegemony.

In turn, this has a two fold importance for the characterization of the events of the 1970s: first, the economic crises can be more readily understood as the results of general contradictions within the maturation of postwar capitalism, evidenced also in an overall shift of US hegemony to an increasingly unilateral and predatory form; and second, the response of the United States to these wider challenges played a significant role in the origins of the oil price increases (which are, therefore, erroneously depicted as the causes of the recessions). It follows that the moment of OPEC was not the decisive rupture with the postwar order that is portrayed by realist and neoclassical commentators, it rather represented the temporary exploitation of a favourable moment in the balance of forces among broader economic and political transformations. And these changes were soon to lead to a diminution of the powers of OPEC, the internal differentiation of the organization, and the subsequent reintegration of its dominant elements – first, Iran and Saudi Arabia, and, second, the Gulf states – into the US economic and strategic orbit. In the domain of oil, therefore, as in the other dimensions of US hegemony, its power was refashioned rather than simply undermined.

The *differentia specifica* of US hegemony – namely, its *capitalist* form and the competitive success of this against the state socialist bloc – is attested to both by the relative failure of the Soviet Union to transform its large resources of petroleum into a comparable international advantage, and by the tutelage exercised by the United States over Western Europe and Japan. This suggests, *contra* realism, that powers cannot simply be ranked by their material resources, for the United States and the Soviet Union are contrasted as Great Powers not merely by their levels of preponderance, but above all by their radically different modes of integration into the world system. Put starkly: capitalist control over oil produced a strategic commodity which served the interests of the hegemon *and* of its allies, and so for the United States this proved to be a positive-sum process; by contrast, communist control resulted in the Soviet Union subsidizing its allies at a huge opportunity cost. In addition, realist expectations fare little better in the domain of military power: for the relative impotence of military force alone is indicated by the inability of the Soviets to exchange military assistance for secure geopolitical support, thus frustrating any attempts to consolidate a radical coalition in the Middle East.[69]

Finally, then, both realism and orthodox economics (whether liberal or mercantilist) have proved unable to register the *qualitative* reorientation of US hegemony noted above, as opposed to its relative, *quantitative* decline as measured by national power-base indices. Although the levels of material preponderance enjoyed by the United States have been somewhat eroded, its focal positions within the world economy and the global military order, as well as its specific geopolitical roles in relation to the Soviet Union and thus to its allies, have more than compensated by extending its influence across the international system.

My conclusions here reinforce the general claim advanced by Susan Strange:

the United States has not in fact lost power in the world market economy. As that economy has grown and spread, the source of its power has shifted from the land and the people into control over the structures of the world system. But the structural power it has acquired in recent decades has been misused in the service of narrow national interests. . . . [During the Reagan administrations] the negative effects of such predatory, or destructive, use of hegemonic power were multiplied by a marked rise in the unilateralism of US policy-making.[70]

If it is also true that in the case of realism or orthodox economics against world oil, which on *prima facie* grounds looked certain to result in the latter's conviction on the basis of diminished responsibility for US hegemony, the defendant has in fact acquitted itself on appeal to a more sober assessment of the evidence, then the first requirement of competent political judgement is that the old jury be excused from all future service.

Postscript August 1990:
Crisis in the Gulf

Early in the morning of 2 August, 1990, Iraq invaded and rapidly overran Kuwait, deposing the ruling al-Sabah family, seizing its domestic assets and taking control of the country's oil installations. Public reaction in the West was nothing if not hysterical. With the prospect of thousands of Western expatriates being held as hostages and Saddam Hussein's proven record of ruthless slaughter, Saddam was compared to Hitler, the situation in the Gulf was likened to that of Europe in the 1930s and the United Nations (UN) was called upon to play its newly allotted role in the 'post-Cold War world' of democracy and peace (a role that the League of Nations had signally failed to play in the thirties). In reality, however, there were few such parallels to be drawn. For while the brutality of Saddam and the aggressive nature of the state capitalist Ba'ath regime is not to be underestimated, the contradictions underlying the conflict revealed a serious crisis of imperialist control over the region and its oil. Beyond the immediate fate of Saddam and the Iraqi Ba'ath, it is the future course of these contradictions, worked out through fast changing indigenous social and national tensions as well as the relations of the Middle East to the capitalist world and the Soviet Union, which will determine the fate of the region.

A long history of imperialist control imposed a series of political and socio-economic structures on the Middle East which were designed to foster Western control over the oil resources of the region. The states of the region were formed largely through the agency of British imperialism (with France playing an important secondary role), both to manage its global strategic interests and to control access to oil reserves. In the postwar period, as the informal empire of American hegemony came to replace the direct administration of Britain and France, strategy centred on integrating the oil-rich economies of the southern Gulf as imperialist relays into a metropolitan circuit of capital and simultaneously arming Israel and Iran as counters against the more populous radical-nationalist Arab regimes of the north. The Gulf producers (in alliance with Iran prior to the Islamic Revolution) thereby disciplined the price of oil for the West in return for military protection, investing their returns in the core rather than sponsoring pan-Arab development. This maintained Western control of the region and its reserves in order to provide a key material base of American hegemony.

Within the Middle East itself, the main threat to that order – from Mossadeq, through Nasser and Khomeini, to Saddam – has come not from the USSR but from indigenous social and national movements seeking to direct resources toward domestic ends, sometimes developmental, more often military and state-building. Strategic management has therefore turned also on isolating or containing any challenge from the more populous states. In addition, through a steady supply of arms transfers and economic assistance, it was hoped that in time nationalist ambitions could be tamed further by integrating local ruling elites into the capitalist world market and hence the Western orbit – in the case of Egypt not without considerable success. Furthermore, while the Soviet Union could arm and aid its allies in the region (Egypt until 1972, Syria, Iraq and the PLO), only the United States could bring direct pressure to bear on Israel in the Palestinian conflict. After the Iranian Revolution resulted in the collapse of the 'twin pillar' strategy of arming Iran and Saudi Arabia as regional clients, the West was prepared to arm and support Iraq's aggression against the new Islamic regime. Throughout, support for Israel continued while the PLO's positions were repeatedly rebuffed. But while Iran was suitably subdued, stability did not return.

Paradoxically, the rivalry of the superpowers and the Cold War also played a stabilizing role, for as long as the Soviet Union remained a significant player in the Middle East it generally sought to temper any ambitions of its principal allies which might embroil it in a direct conflict with the United States, while taking opportunistic advantages where possible. But the Soviet position was always weak in the oil industry – compare the position of either Iraq or Syria with that of the Gulf states and Saudi Arabia – and its relations have recently deteriorated vis-a-vis its allies as it has turned its attention to concentrate on internal issues and is now quite unprepared to risk Western antagonism. Once the Soviet Union began to retreat, its main ally, Iraq, had a freer hand. Moreover, the block to Western dominance of the UN provided by the

Soviet veto on the Security Council, together with the votes of its allies, has all but evaporated.

Overall, this postwar structure of control embodied a number of potentially explosive contradictions. Most obviously, the United States was the economic and military backer of both the conservative Islamic monarchies of the Gulf and the state of Israel. But more important, materially speaking, were three aspects of the socio-economic development of the Middle East. First, the Gulf states had in effect exchanged US military protection in return for managing the region's oil in the interests of the Western consuming nations rather than the rest of the Arab world. Second, the class basis of the nationalist regimes, together with the willingness of both superpowers to arm local clients, provided a fertile soil for authoritarian and highly repressive regimes. And third, the integration of local ruling classes into the capitalist world, combined with the essentially domestic base of their social and material reproduction, aggravated indigenous ethnic, religious and social tensions. Given these deep social tensions and long-standing nationalist hostility to the contrived and unstable political order in the region, conflict was always a possibility. For although hegemonic powers can organize economies and protect states, they cannot for ever contain the disruptive effects of social and national revolt, as the latest conflict attests. It is the (temporary?) loss of this control which has made the current crisis possible, and which has inevitably called forth the forceful US actions. At the same time, Soviet weakness has provided the basis for a fragile unanimity in the UN and has also freed the US capacity for intervention.

United States hegemony during the postwar era has been and remains materially dependent upon a directive role in the international oil industry. Specifically, after the majors lost direct control in the mid-1970s, this role came to depend on the connections established to the Gulf producers and especially to Saudi Arabia. And throughout the later 1980s the United States provided strong support for Iraq's aggression against Iran. Regional stability and the integrity of the Gulf is simply a vital basis for US power in the international system. An important but strictly subsidiary issue concerns the rising oil imports into the United States itself. (Since 1985 US oil imports have risen from 31 to 52 per cent of domestic consumption, with oil from the Gulf increasing from 5 to 24 per cent. Over the same period, US domestic production fell by 15 per cent. This is not surprising given that nearly two-thirds of US oil consumption goes on transport, and that the failure to tax petrol consumption has left prices more than 50 per cent lower than in, say, Britain, halving the real cost since the oil crisis.) The EC and Japan have followed the US lead both because of their much greater dependence on the region for oil supplies and because they remain tied to US leadership in world affairs in a number of ways. These dynamics can be traced in the unfolding of the crisis itself.

Saddam's Gamble

It is too early to give anything like a complete explanation of these momentous and dangerous events. But if the crisis represents both a loss of imperialist control and fierce intra-Arab conflict, what specifically lay behind this unexpected aggression and the rapid and US-sponsored international response? The immediate cause of Saddam's actions was a dispute within OPEC over pricing and production policy. The realignments within OPEC during the 1980s, which involved a Gulf-led drive to moderate prices and regain market share, placed considerable economic pressure on Iraq and its economic straits were indeed dire. Had OPEC's 1986 reference price of $18/barrel kept pace with inflation, crude oil would have stood at $22.7/barrel by the summer of 1990 – in fact, prices wavered between $17–18/barrel. The Gulf producers had supported the real price fall in order both to regain the market share that OPEC had lost in the first half of the 1980s and to slow down the transition away from oil to other fuels as well as the development of new forms of supply. In fact, Kuwait and the United Arab Emirates were the main quota 'cheats' within OPEC. On the other hand, Iraq (along with Iran) was desperate for foreign exchange and state revenues in order to undertake urgent postwar reconstruction. Iraq had some $65 bn of outstanding debt of which perhaps $30 bn was owed to the West; and Iraq's oil exports in 1989 at 2.8 mb/d earned $12 bn or virtually all of its export earnings, accounting for three-quarters of GNP and most of the finance for the state. So, after some 'sabre-rattling' along the Kuwaiti border by Iraq, the OPEC Conference on 26 July aimed for a reference price of $21/barrel with a production ceiling of 22.5 mb/d for the second half of 1990 (production had reached 24 mb/d as a result of cheating).

More generally, Iraq's main demands on Kuwait and the southern Gulf were:

1 strict obedience within OPEC to the quotas necessary to sustain the oil price;
2a a claim of $2.4 bn on Kuwait for oil it alleged had been 'stolen' from the Rumaila field,

b an Arab 'Marshall Plan' to rebuild the Iraqi economy, a figure of $14 bn was mooted, and

c the write-off of the $35 bn of war loans from the Gulf states, including some $10 bn from Kuwait; and

3 territorial claims involving the Kuwaiti portion of the Rumaila oil field as well as access to Kuwait's Warbah and Bubiyan islands near the Faw peninsula in the Gulf.

With the price issue for the moment resolved, Saudi Arabia sponsored talks between Iraq and Kuwait on the oustanding issues which broke up on 1 August without progress, Kuwait refusing simply to capitulate to Saddam's demands. The invasion began the next day. And by 8 August, a puppet Iraqi regime and defence force had matured into an outright annexation. By controlling Kuwait's oil Saddam virtually doubled Iraq's reserves to nearly 196 bn barrels, gained potential foreign assets of perhaps $100 bn, and secured a potential presence in the oil market approaching that of Saudi Arabia. If successful, the long-term economic status of Iraq would be commensurate to that of Saudi Arabia and the other Gulf states, which together with its population and military forces would have brought it strategic dominance in the region as well as posing a direct challenge to the imperialist control of the Gulf's oil. It was this combined threat which the United States simply could not allow to stand.

Now, while these economic and strategic incentives must have weighed heavily in Saddam's calculations, other factors were at work. Notwithstanding the PLO's formal acceptance of a 'two state' solution to the Palestinian problem in 1988, the prospects for meaningful negotiations seemed as distant as ever. And with the Soviet Union playing a reduced international role in general, and offering less support to its Middle East allies, Iraq, Syria and the PLO, in particular, many Arabs felt increasing frustration, especially given the inability or unwillingness of the United States to impose serious pressure on Israel to negotiate. Saddam's drive for Arab leadership had therefore come to command considerable respect as well as fear among many Arabs outside the Gulf itself. The most important Arab allies of the United States in the region, Saudi Arabia and the other Gulf states, Egypt and Jordan, were of course hostile to the line pursued by Iraq but could not be seen in open and explicit agreement with Western interests. For while the fortunes of the ruling classes of the 'moderate' Arab regimes were linked to continuing integration into the world market as well as more or less directly to US economic and military assistance, the support of the United States for Israel combined with the domestic insecurity –

both political and socio-economic – of most Arab regimes demanded the maintenance of at least the semblance of Arab unity.

Furthermore, the consequences of oil wealth, the Iran–Iraq war and plentiful arms supplies had left Iraq with a formidable military capacity, many times greater than that of all the Arab Gulf states put together for example. During the 1980s, it is estimated that Iraq spent some $80 bn on weapons, with the major supplies coming from the Soviet Union (53 per cent), France (20 per cent) and China (7 per cent). This alone is enough to demonstrate the self-serving character of Saddam's attempt to lead the Arab world and his own lack of concern with regional development. Iraq's state capitalism and the ideology of Ba'ath rule amount to a ferociously repressive form of dictatorship, with little concern for pan-Arab development. Still, with this awesome military capacity and a proven record of ruthless deployment. Saddam could count on a suitably subdued response from his Arab neighbours.

Nevertheless, there was considerable popular antagonism directed at the huge wealth of the Gulf states among the Arab population as a whole. The oil-backed pattern of regional development has left a poor and underdeveloped majority looking at the wealthiest societies in the world. The anomaly of the Gulf states, holding most of the region's oil but a minority of its population, was widely resented. Kuwait, with the highest GDP per capita in the world and more expatriate labour than native Kuwaitis, provides an exemplary illustration. In 1989 Kuwait earned more income from its vast overseas investments located in the capitalist metropoles than from its domestic oil production ($8.8 bn as compared with $7.7 bn), invested only some 15 per cent of its state reserves in other Arab economies, and had a labour force composed of native Kuwaitis (18 per cent), Asians (42 per cent), Arabs (38 per cent), and Westerners (2 per cent). An extreme example to be sure, but none the less a dramatic illustration of the lack of economic integration in the Arab world. Here too Saddam could expect to tap deep wells of support, finding a ready reception among the Arab masses for his anti-imperialist rhetoric.

Finally, the Bush administration had been slow to adapt to the pro-Iraqi, anti-Iranian/Syrian stance it inherited from the Reagan team. US–Iraqi ties strengthened in the closing stages of the Iran–Iraq war. (At that time, the much-vaunted 'international community' kept Saddam supplied with weapons and intelligence, while the United States all but entered the war on Iraq's side with its attacks on Iran's oil installations and policing of the Gulf. Moreover, the UN stood by and did nothing while Saddam deployed chemical weapons to slaughter the Kurds.) In addition, since the July 1988 cease-fire in the war, Iraq has been the main obstacle to a lasting peace settlement, retaining many Iranian prisoners and refusing to accept the border settlement and the division of the Shatt al-Arab waterway agreed to in the Algiers Agreement of 1975. The West had put no pressure on Iraq to compromise. This policy of benign neglect towards Iraq may have encouraged Saddam to believe that his actions would go if not unchallenged then at least unpunished.

If the lure of Kuwaiti wealth and, with it, regional dominance provided the attraction, and if regional and international circumstances presented the opportunity, then driving the whole was a profound state of tension and socio-economic crisis within the Iraqi regime itself. Saddam's Ba'ath regime, comprising a narrow social base with a highly centralized and brutally repressive political order, straddles a society of considerable ethnic and religious diversity. (The regime has been compared both to European fascism and Stalinism.) The huge costs of the war against Iran had resulted in the cancellation or delay of much needed economic development, and the anticipated postwar reconstruction has been disappointing. The crisis was clearly spreading into the state apparatus itself, as the purges of the Ba'ath Party and the recurrent executions of senior military functionaries attested. Under these circumstances, Saddam appeared to have staked his survival on an aggressive policy of annexation to garner oil assets and bolster support throughout the (non-Gulf) Arab world. Perhaps he also calculated that the internal insecurity of the pro-Western Arab regimes, combined with the support of the United States for Israel, would preclude outside intervention. The annexation of Kuwait might then become simply a regrettable but acceptable *fait accompli*. Or perhaps this was a desperate gamble by a crisis-torn regime.

The Crisis Unfolds

The opening international response to Iraq's belligerence was a near-universal condemnation of the invasion, followed by, on the one hand, a rapid convening of the UN Security Council, which called for an unconditional return to the status quo *ante bellum*, and, on the other, a joint US-Soviet condemnation of Iraq's actions. The Soviet Union, a longstanding ally of Iraq and its principal military supplier, adopted a much more low-key approach but nevertheless reluctantly co-operated in the UN and issued a form of joint statement with the United States. In the Arab world itself, the picture was altogether more complicated. The Arab League condemned Iraq (although Jordan, Libya, the PLO,

Yemen, Mauritania and Sudan abstained), as did the Gulf Co-operation Council (GCC) and the Islamic Conference Organization.

But whilst continuing and confused intra-Arab attempts at mediation foundered amongst Arab disunity and considerable uncertainty as to Saddam's intentions, within days a coherent and decisive response had been fashioned under strong US leadership. There were two basic components of the overall strategy. In the first place, an economic embargo and boycott was to be imposed on Iraq and Kuwait. Secondly, US military power would be committed to the region to guarantee the stability of the Persian Gulf and, above all, the integrity of Saudi Arabia. On 12 August, US forces indicated that they would interdict all trade to and from Iraqi-controlled territory. The basis for this action was Kuwait's call for collective self-defence under Article 51 of the UN Charter. To be effective it would be necessary (1) to prevent the export of Iraqi and Kuwaiti oil, (2) to deny Iraq much needed imports, (3) to restabilize the oil market, and (4) to pre-empt and deter any Iraqi aggression against Saudi Arabia and, especially important, prevent the destruction of Saudi oil facilities. The aims of this strategy were to bleed the already-pressed Iraqi economy dry, thus provoking internal disunity and spreading disaffection among the population, the Ba'ath Party and the army, with the hope of toppling Saddam and installing a more conciliatory leadership. Meantime, the accumulating US military power would police the blockade, serve to defend Saudi Arabia, and (possibly) act as a means of direct attack on Iraq's forces and key infrastructure.

Iraq's export earnings were almost entirely derived from the 2.7 mb/d of oil exported, of which 1.78 mb/d went to the United States, Japan, the EC and Turkey, while nearly one half of Kuwait's 1.7 mb/d had gone to the EC, Japan and the United States. This was boycotted almost immediately, as the United States and the EC froze Kuwaiti and Iraqi assets, boycotted trade and investment, stopped arms sales and, most important of all, banned all oil imports from Iraqi-occupied territory. Japan and Turkey followed suit after strong US diplomatic pressure. Moreover, some two-thirds of Iraq's imports also came from the EC, the United States, Turkey and Japan. (Iraq's vulnerability was all the greater in that 70–80 per cent of its food is imported.) But as in the case of other attempts to impose economic sanctions, substitute customers might be available. Iraqi oil could leave by four routes: by pipeline across Turkey to ports in the Mediterranean; by pipeline across Saudi Arabia to the port of Yanbu on the Red Sea; from ports in the Gulf; and by truck across to Turkish ports and the Jordanian port of Aqaba, also on the Red Sea. All of Kuwait's exports left from Gulf ports. (There is also a pipeline for Iraqi oil across Syria to the eastern Mediterranean. This was closed by Syria when it sided with Iran in the Iran–Iraq war and remained shut at the outset of the crisis.) So a complete isolation demanded the closure of the pipelines and a naval blockade of the Gulf, the Red Sea and the eastern Mediterranean to interdict the 2 mb/d of Iraqi-controlled capacity not already boycotted.

In order to enlist support, the Bush administration undertook considerable diplomatic effort to ensure that the action taken was seen as 'international' and not simply American or even Western. The British Prime Minister, Thatcher, also played an active role in urging the international community, and particularly the NATO allies and the member states of the European Community (EC), to support the US lead. The Soviet Union and China suspended shipments of arms to Iraq. On 6 August the UN Security Council voted 13–0 (with abstentions from Cuba and Yemen) to impose a complete embargo on and a boycott of economic links with Iraq and Kuwait. (This also opened the possibility of a UN military blockade, but this would require further action by the Security Council.) On 9 August the Security Council repudiated Iraq's formal annexation of Kuwait. Turkey, both a member of NATO and a close and long-standing ally of the United States, closed its pipeline in accordance with the UN Resolution on 7 August. (On 14 August, Turkey announced that Kuwait had agreed to pay compensation for the economic costs of sanctions – Turkey exported considerable amounts of food to, and received some 60 per cent of its oil from, Iraq.) Turkey has consistently received generous amounts of economic assistance and military aid and co-operation from the United States, in part because of its long border with the Soviet Union but also as a result of its proximity to the oil reserves of the Middle East. Thus, despite some domestic criticism of President Turgut Ozal's response, co-operation with the West was assured given Turkey's long-term ambition to join the EC and its need for a post-Cold War status in NATO.

Also on 7 August – a mere five days after the invasion – US forces were deployed inside Saudi Arabia, principally in the eastern oil-rich region around Dhahran, close to the border with Iraqi-occupied Kuwait. Explained in terms of an imminent Iraqi attack on the Kingdom, possibly involving chemical weapons, the Saudi acquiescence to US strategy was also reported to involve a promise to close the Iraqi pipeline to Yanbu and to increase Saudi oil production. It soon became clear that as well as defensive forces the United States was amassing a huge strike force capable of initiating direct attacks against Iraq. Moreover, although described as 'international' by the United States and

Britain, the growing military task force in the eastern Mediterranean, the Red Sea and the Gulf was not initiated under either UN or NATO auspices. This was a response by the United States acting in concert with NATO allies (especially, Britain and Turkey) and guarding its key regional protectorate, Saudi Arabia. The legal bases of the action were requests for assistance from Saudi Arabia and Kuwait. Indeed, by 13 August, all military facilities belonging to the states of the Gulf Co-operation Council (Saudi Arabia, Oman, Bahrain, UAE, Qatar and Kuwait) were at the disposal of US and UK forces. However, the United States did succeed in gaining strong backing from later meetings of NATO and EC foreign ministers as well as some military co-operation. Also a more active UN role might still emerge.

In the Middle East, the situation remained confused. The United States sought to encourage an Arab presence in Saudi Arabia, especially from Egypt and Morocco. On 10 August the Arab League voted 12–8 (with opposition from Iraq, Libya and the PLO) both to endorse the positions of the UN Security Council and to send an Arab force to assist in the defence of Saudi Arabia, with Egypt, Syria and Morocco participating. On 13 August Pakistan also announced its intention to contribute to the defence of Saudi Arabia. Jordan also announced its intention to comply with the UN sanctions but has yet to take action. With over one million Palestinian inhabitants, much of its trade destined for Iraq, and 95 per cent of its oil on easy credits from Iraq, the position of Jordan (a strong pro-Western influence) and the survival of King Hussein was precarious. Nevertheless, the United States made plain that a blockade of the Jordanian port of Aqaba had not been ruled out. Meanwhile, Saddam called for an Arab uprising against the US presence in Saudi Arabia and the Gulf and for the Arab people to overthrow their 'corrupt' pro-Western leaders. Pro-Iraqi rallies, bringing together pan-Arab nationalists and Islamic militants, were reported to have taken place in Algeria, Tunisia, Libya, Yemen, Jordan, the Israeli-occupied territories and criticism was registered in Syria and Egypt. Also after Saddam's threats towards Israel of 8 August the Israeli leadership issued strong and unmistakeable warnings on 9 August. It was also reported that Iran had offered to defend the UAE, Oman, Bahrain and Qatar. (On 15 August, Saddam made conciliatory noises towards Iran in respect of the peace settlement in order to increase his room for manoeuvre elsewhere.)

Oil prices rose rapidly in the early stages of the crisis. Brent Crude, for example, increased from around $18.75/barrel in early July, through $21.5/barrel after the 26 July OPEC meeting, to $26.5/barrel on 7 August. But with a growing expectation of compensatory increases in production, together with the meeting of the International Energy Agency (IEA) on 9 August to discuss the possible release of reserves, prices fell back to around $25/barrel. The IEA called on the majors to use their large stocks and hinted that it was ready to release public supplies when and if necessary. Conciliatory indications also came from OPEC. A restabilization of the oil markets was necessary so that the removal of Iraqi and Kuwaiti production did not drive prices upwards, thereby precipitating instability in the world economy and undermining support for US action. (Both the OECD and OPEC states had a strong interest in such stability.) Past precautionary measures and the slack market meant that enough spare capacity existed in the system to meet a maximum shortfall of 4.5 mb/d. Saudi Arabian production could be increased by around 2 mb/d and that of the UAE by 700,000 b/d, Venezuelan output could be raised by 500,000 b/d and both Nigeria and Iran could also increase exports. But even without the co-operation of these OPEC states (and Saddam made clear that he regarded such as an act of aggression), or if they refused assistance until the West's own oil stocks had been somewhat depleted, the OECD's public reserves alone (excluding large private company reserves) amounted to an estimated 30 days' consumption. This implied that at the rate necessary to replace Iraqi and Kuwaiti output, public OECD supplies would last for perhaps two years – significantly longer than Saddam or Iraq could survive economic isolation. If properly co-ordinated and clearly signalled to the market, there was no medium-term danger of a price explosion. This was also the message publicly transmitted by the majors.

Prospects

As noted above, while the United States strenuously sought to inter-nationalize the response to the crisis, with some considerable success in the UN and in relations with its NATO partners, the response was in fact primarily lead, orchestrated and directed by the United States itself. Thus, by 10 August the Soviet Union was warning that unilateral action might escalate the situation and suggested that it might partici-pate militarily in a UN-sponsored force (and around 1,000 Soviet military advisers remained in Iraq). On the other hand, Iraq sought to portray the response of the United States as an imperialist and 'Zionist' plot against the 'Arab nation'. Now, while the talk of an 'Arab nation' has precious little genuine social or material foundation, the charge of imperialist control has considerable cogency and struck a resonant cord throughout the Arab world. At the time of writing (15 August), the situation is as unpredictable as it is dangerous. Iraq may be cowed by its economic isolation and Saddam deposed by elements in the Ba'ath Party and the army seeking to save themselves; Saddam might escalate the situation either by attacking Saudi Arabia or Israel, desperately seeking to fashion an 'anti-imperialist' unity among the Arabs; some Arab-negotiated face-saving formula might yet be created; the United States might seek a pretext to launch direct attacks on Iraqi territory; or the crisis could spread to the world economy undermining Western resolve. (The cost of the US operation, before any fighting has taken place, is already estimated at between $300–400 million per month, placing considerable pressure on Bush to find a rapid resolution.)

But whatever the outcome of the current crisis, the plain fact remains that the murderous regime crafted by Saddam sits in a cauldron of the West's making. For the West's policy, the state structures of the area, and its pattern of socio-economic development were fashioned to maintain imperialist control, British and then US, of the region and its oil. Whatever the fate of Saddam and the Ba'ath, this suggests that nothing short of profound socio-economic and political change – both within the region and in its relations with the capitalist metropoles – can guarantee against a future upheaval which would once again bring the Middle East to the brink of disaster. If all goes well for the United States and Saddam falls, then perhaps the West will be tempted to try to impose a 'friendly' regime, gaining even further control over the region's oil. But long-term peace and stability are unlikely to be secured without four basic alterations in the Western posture; a formula must be devised to stabilize movements in the oil price, perhaps by linking it to that of industrial goods in the core; a more auto-centric pattern of regional development needs to be developed, spreading Gulf wealth to the poorer Arabs of the north rather than the markets of the West; the regional build-up of armaments must be curtailed and a lasting security structure must replace the balance of terror previously provided by superpower competition, economic aid replacing military credits; and a solution must be negotiated to the Palestinian question. Sadly, these kinds of change appear as distant and difficult as ever.

In the longer-run still, the recent events have placed a large question-mark over a range of issues. Will the United States and the West be able to fashion a new form of control which will guarantee stable access to the region's oil at relatively cheap prices and, if not, what then? What is to become of the emerging regional powers such as Turkey and Egypt, what role will they play? And, given the much greater oil-dependence of Japan and the EC what do these events portend for Japan's self-denying ordinance not to commit forces except in self-defence and for NATO's rethinking of its post-Cold War, out-of-area role?

Notes

INTRODUCTION: THE BASIC ARGUMENT

[1] See for example, Robert E. Gilpin (with the assistance of Jean M. Gilpin), *The Political Economy of International Relations*, Princeton University Press, Princeton 1987; or from a very different perspective, John Palmer, *Europe without America?*, Oxford University Press, Oxford 1987.

[2] See, for example, Robert O. Keohane, *After Hegemony*, Princeton University Press, Princeton 1984.

[3] Susan Strange, *States and Markets*, Pinter, London 1988. While finding Strange's insistence on the requirement for international analysis to pay attention to the ramifications of structural power refreshing, as can be seen in what follows I do not agree that the structures she has chosen are at all appropriate for the task in hand.

[4] Stephen D. Krasner, *Structural Conflict*, University of California Press, Berkeley 1985.

CHAPTER 1 RETHINKING INTERNATIONAL RELATIONS THEORY

[1] Cf. Susan Strange's comment that 'the study of international relations has tended in many places to divide into "strategic studies" on the one hand and the "politics of international economic relations" on the other' (*States and Markets*, Pinter, London 1988, p. 46).

[2] Steve Smith has appropriately noted that 'the division of International Relations from other social sciences seems more a logical consequence of the pervasiveness of Realist assumptions than a reflection of some specific features of its subject matter' ('Paradigm dominance in international relations: The development of international relations as a social science' in *Millennium: Journal of International Studies*, 16, p. 199).

[3] This debate was sparked off by Kenneth Waltz (*Theory of International Politics*, Addison-Wesley, Reading, Mass. 1979) and can be sampled in

Robert O. Keohane (ed.), *Neorealism and its Critics* Columbia University Press, New York 1986.

4   See, for example, E. A. Brett, *The World Economy since the War* Macmillan, London 1985; Stuart Corbridge, *Capitalist World Development* Macmillan, London 1986; and Chris Edwards, *The Fragmented World* Methuen, London 1985.

5   This debate is admirably summarised in Andrew Webster, *Introduction to the Sociology of Development* Macmillan, London 1984.

6   See, for example, Peter B. Evans, Dietrich Rueschemeyer and Theda Skocpol (eds), *Bringing the State Back In* Cambridge University Press, Cambridge 1985.

7   The work of Michael Mann is particularly relevant in this context: 'Capitalism and Militarism', in Martin Shaw (ed.), *War, State and Society*, Macmillan, London 1984; 'The roots and contradictions of modern militarism' in *New Left Review*, no. 162; 1987, pp. 35–50 and 'War and social theory: Into battle with classes, nations and states', in Colin Creighton and Martin Shaw (eds), *The Sociology of War and Peace*, Macmillan, London 1987.

8   See Michael Mann, 'The Autonomous Power of the State: Its Origins Mechanisms and Results', originally in *Archives europeennes de sociologie*, 25, 1984 and reprinted in John Hall (ed.), *States in History*, Basil Blackwell, Oxford 1986 from which references are taken; and Anthony Giddens, *The Nation-State and Violence*, Polity Press, Cambridge 1985.

9   From a large literature, see especially the works of Roy Bhaskar (*The Possibility of Naturalism* Harvester Press, Brighton 1979 and *Scientific Realism and Human Emancipation,* Verso, London 1986) and of Anthony Giddens (*Central Problems in Social Theory*, Macmillan, London 1979 and *The Constitution of Society*, Polity Press, Cambridge 1984).

10 Steve Smith provides such an account, without, however, assenting to the cognitive claims made for realism ('International relations', in Lynton Robbins (ed.), *Introducing Political Science*, Longman, London 1985). For a cogent debunking of classical realist pretensions, see Justin Rosenberg, What's the Matter with Realism?, in *Review of International Studies*, forthcoming 1990.

11 Edward Hallett Carr, *The Twenty Years' Crisis, 1919–1939*, Macmillan, London 1939 and Hans J. Morgenthau, *Politics among Nations* Alfred A. Knopf, New York (6th edn, with Kenneth W. Thompson), 1985.

12 Carr, *The Twenty Years' Crisis*, p. 15.

13 Morgenthau, *Politics among Nations*, p. 3.

14 Ibid., p. 4.

15 Ibid., p. 5.

16 Steve Smith has noted that postwar (one might even say, Cold War) realism has been very much an American affair: 'Only in the US has there been the combination of an intellectual predisposition towards social science, a system of policy communities that takes peoples back and forth from the academic and political worlds, and a political climate that was looking for

guidelines for managing international events' ('Paradigm dominance in international relations', p. 203).

17 On the radical and Marxian left, of course, political economy never fell out of favour, but in polite circles Richard Cooper's *The Economics of Interdependence* (McGraw-Hill, New York 1968) paved the way for its rediscovery. See Strange, *States and Markets*, pp. 20–1.

18 See the symposium edited by Stephen D. Krasner, *International Regimes*, Cornell University Press, Ithaca 1983.

19 See in particular the magisterial survey by Robert Gilpin (with the assistance of Jean M. Gilpin), *The Political Economy of International Relations*, Princeton University Press, Princeton, 1987.

20 For example, Robert Gilpin, *War and Change in World Politics*, Cambridge University Press, Cambridge 1981; and *The Political Economy of International Relations*.

21 Waltz, *Theory of International Politics*.

22 See, for example, Richard K. Ashley, 'The Poverty of Neorealism' in Keohane (ed.), *Neorealism and its Critics*.

23 One instance is the collection edited by Michael Smith, Richard Little and Michael Shackleton, *Perspectives on World Politics*, Croom Helm, London 1981.

24 Robert O. Keohane and Joseph S. Nye, *Power and Interdependence*, Little, Brown and Company, Boston 1977.

25 Ibid., pp. 23–37.

26 Ibid., p. 25.

27 Ibid., pp. 29–30.

28 Ibid., p. 38.

29 See, for example, Roger Backhouse, *A History of Modern Economic Analysis*, Basil Blackwell, Oxford 1985, pp. 123–7; I take this last felicitous phrase from the important essay by Anthony Giddens, 'Social theory and problems of macroeconomics' in *Social Theory and Modern Sociology*, Polity Press, Cambridge 1987, p. 187.

30 Eric J. Hobsbawm, *The Age of Empire, 1875–1914*, Weidenfeld and Nicolson, London 1987, p. 270.

31 For a simple, formal exposition of general equilibrium theory, see E. Roy Weintraub, *General Equilibrium Theory*, Macmillan, London 1974. For two accessible surveys of general neo classical theory which contrast it with the surplus tradition, see M. C. Howard, *Modern Theories of Income Distribution*, Macmillan, London 1979, and *Profits in Economic Theory*, Macmillan, London 1983. See also Vivian Walsh and Harvey Gram's important comparative treatment, *Classical and Neoclassical Theories of General Equilibrium*, Oxford University Press, Oxford 1980; and the work of Stephen A. Marglin, *Growth, Distribution, and Prices*, Harvard University Press, Cambridge, Mass. 1984. And for two important historical treatments which provide clear accounts of the differences involved, see Maurice Dobb, *Theories of Value and Distribution since Adam Smith*, Cambridge

University Press, Cambridge 1973; and Phyllis Deane, *The Evolution of Economic Ideas*, Cambridge University Press, Cambridge 1978.

32 Lionel Robbins, *An Essay on the Nature and Significance of Economic Science*, Macmillan, London 1932, p. 16.

33 For detailed discussion, see John Eatwell, 'Theories of value, output and employment' (*Thames Papers in Political Economy*, London 1979, cited from a revised version which appears in J. Eatwell and M. Milgate (eds), *Keynes's Economics and the Theory of Value and Distribution*, Duckworth, London 1983), and especially the excellent methodological discussion in Sheila C. Dow, *Macroeconomic Thought*, Basil Blackwell, Oxford 1985.

34 See Geoff Hodgson, 'Behind methodological individualism', in *Cambridge Journal of Economics*, 10, 1986, pp. 211–24.

35 See Frank Hahn's Introduction, 'On the notion of equilibrium in economics' and 'General equilibrium theory' in idem, *Equilibrium and Macroeconomics*, Basil Blackwell, Oxford 1984. In the early 1980s, Hahn memorably compared general equilibrium theory to:

> the mock-up an aircraft engineer might build. My amazement in recent years has accordingly been very great to find that many economists are passing the mock-up off as an airworthy plane, and that politicians, bankers and commentators are scrambling to get seats. This at a time when theorists all over the world have become aware that anything based on this mock-up is unlikely to fly, since it neglects some crucial aspects of the world, the recognition of which will force some drastic redesigning. Moreover at no stage was the mock-up complete; in particular it provided no account of the actual working of the Invisible Hand. (Quoted in Keith Smith, *The British Economic Crisis*, Penguin, Harmondsworth 1984, pp. 148–9.)

36 Hodgson, 'Behind methodological individualism', pp. 221–2.

37 Ibid., p. 220.

38 Eatwell, 'Theories of value, output and employment', p. 94.

39 For a standard, realist account of Marxism's limitations as a theory of international relations, see the confused and tendentious study by Vendulka Kubalkova and Albert Cruickshank, *Marxism and International Relations*, Oxford University Press, Oxford 1989; and for a rare assessment of Marx's writings on international issues, see R. S. Neale, 'Marx and Lenin on imperialism', in *Writing Marxist History*, Basil Blackwell, Oxford 1985.

40 It is a measure of the lack of attention paid to Marxist arguments by mainstream international relations theory that even where these debates are considered Lenin's popular outline provides the principal reference point (Vladimir I. Lenin, *Imperialism, The Highest Stage of Capitalism*. It is also worth noting that Lenin's original title was 'Imperialism, the *Latest* Stage of Capitalism' (my emphasis); it was only the posthumous editions which were retitled – see Eric Hobsbawm, *The Age of Empire, 1875–1914*, pp. 12 and 363, n. 3). See, for example, Gilpin, *The Political Economy of International Relations*. For the political and economic context of Bukharin's writings, see Stephen F. Cohen, *Bukharin and the Bolshevik Revolution*, Oxford University Press, Oxford 1980, pp. 25–43. More generally, Anthony

Brewer concludes his excellent review of the classical debate on imperialism thus:

> Bukharin combined the analysis of the *internationalisation* of capitalist relations of production (Marx, Luxemburg) with Hilferding's analysis of the formation of blocs of finance capital to show why these blocs formed on a *national* basis. The competitive struggle continues, in the era of finance capital, but it now takes the form of military and political rivalry between "state capitalist trusts". Lenin's *Imperialism* follows the same lines on a lower level of abstraction, providing a forceful descriptive account of imperialism'. (*Marxist Theories of Imperialism*, Routledge & Kegan Paul, London 1980, p. 126.)

See also Fred Halliday, 'Vigilantism in international relations: Kubalkova and Cruickshank and Marxist theory', in *Review of International Studies*, 13, 1987, pp. 215–29.

[41] For this reading of Lenin and for an important discussion of his work, see Diane Elson, 'Imperialism', in Gregor McLennan, David Held and Stuart Hall (eds), *The Idea of the Modern State*, Open University Press, Milton Keynes 1984. I take up this discussion in more substantive detail in chapter 2.

[42] A point well made by Gilpin, *The Political Economy of International Relations*, pp. 34–41.

[43] Fred Halliday, 'Vigilantism in international relations', p. 171.

[44] Ernest Mandel, *Trotsky*, New Left Books, London 1979, p. 34.

[45] The contemporary use of these analyses can be seen most clearly in the work of Mike Davis, 'Nuclear imperialism and extended deterrence', in E. P. Thompson, *Exterminism and Cold War*, Verso, London 1982; and in a significantly modified form in Fred Halliday, *The Making of the Second Cold War*, Verso, London, 1983.

[46] See Simon Clarke, *Keynesian Monetarism and the Crisis of the State*, Edward Elgar, Aldershot 1988, p. 16, n. 19; and for other discussions which appreciate the importance of this central theme of Marx's work, see also Makoto Itoh, *Value and Crisis*, Pluto Press, London 1980; John Weeks, 'Equilibrium, uneven development and the tendency of the rate of profit to fall', in *Capital and Class*, no. 16, 1982, pp. 62–77; and Ben Fine, *Marx's Capital*, Macmillan, London 1989, pp. 59–74. More general treatments include, Meghnad Desai, *Marxian Economics*, Basil Blackwell, Oxford 1979; Ben Fine and Laurence Harris, *Rereading Capital*, Macmillan, London 1978; Geoff Hodgson, *Capitalism Value and Exploitation*, Martin Robertson, Oxford 1982; Bob Rowthorn, 'Neo-classicism, neo-Ricardianism and Marxism', in *Capitalism, Conflict and Inflation*, Lawrence and Wishart, London 1980; and John Weeks, *Capital and Exploitation*, Edward Arnold, London 1981.

[47] Richard Walker, 'The dynamics of value, price and profit', in *Capital and Class*, no. 35, 1988, pp. 146–81; and R. M. Goodwin, 'Swinging along the turnpike with von Neumann and Sraffa', in *Cambridge Journal of*

Economics, 10, 1986, pp. 203–10; see also Eatwell, 'Theories of value, output and employment'.

48 These are points well made by Walker, 'The dynamics of value, price and profit'.

49 Ibid., p. 171.

50 Quoted in Gilpin, *The Political Economy of International Relations*, p. 32. However, for the real history of mercantilism, see Maurice Dobb, *Studies in the Development of Capitalism*, Routledge and Kegan Paul, London 1946, pp. 177–220.

51 Gautam Sen, *The Military Origins of Industrialisation and International Trade Rivalry*, Frances Pinter, London 1984.

52 Again a point well made by Gilpin, *The Political Economy of International Relations*, p. 42.

53 See also Gilpin, *War and Change in World Politics*, pp. 93–4.

54 The latter point is completely overlooked in Waltz's critical discussion of Marxist theories of imperialism, a characteristic error which vitiates most of his argument on this point. See *Theory of International Politics*, pp. 18–37.

55 Michael Barratt Brown, *The Economics of Imperialism*, Penguin, Harmondsworth 1974, p. 68.

56 Stephen D. Krasner, 'Structural causes and regime consequences: regimes as intervening variables', in idem. (ed.), *International Regimes*, p. 1; see also Robert O. Keohane in *After Hegemony*, Princeton University Press, Princeton 1984; and for a powerful critique of the regimes school, see Susan Strange, '*Cave! hic dragones*: a critique of regime analysis', in Krasner (ed).

57 Krasner, 'Structural causes and regime consequences', p. 9.

58 Stephen D. Krasner, 'Regimes and the limits of realism: regimes as autonomous variables', in idem. (ed.), *International Regimes*, pp. 367–8.

59 Waltz, *Theory of International Politics*, pp. 5, 6, 8 and 13.

60 See Anthony Giddens, 'Classical theory and modern sociology', in *American Journal of Sociology*, 1976 (reprinted in idem., *Profiles and Critiques in Social Theory*, Macmillan, London 1982).

61 Thus Waltz writes that: 'One cannot infer the conditions of international politics from the internal composition of states, nor can one arrive at an understanding of international politics by summing the foreign policies and external behaviors of states', *Theory of International Politics*, p. 64. And while elaborating this distinction between reductionist and systemic theories, Waltz is careful not to confuse it with another: namely, the reductionist explanation of one system by causes deriving from another (such as Wallerstein's alledged reduction of the nation-state system to an effect of the capitalist world economy), see ibid., pp. 38–9.

62 Waltz, *Theory of International Politics*, pp. 71–2.

63 Ibid., p. 69.

64 While the Copernican Revolution may have 'turned upon the most obscure

and recondite of astronomical minutiae', it fundamentally transformed our basic social and scientific self-understanding; it is not immediately apparent how Waltz's parallel achievement might have similar effects. See Thomas S. Kuhn, *The Copernican Revolution*, Harvard University Press, Cambridge, Mass. 1957, p. 1.

65 Waltz, *Theory of International Politics*, pp. 79–101.

66 Ashley, 'The poverty of neorealism', p. 271. Ashley also rightly notes that: 'In neorealism, there is no concept of social power behind or constitutive of states and their interests. Rather, power is generally regarded in terms of capabilities that are said to be distributed, possessed, and potentially used *among* states-as-actors' (p. 276). It is a sad reflection on the level of theoretical literacy among realists that both Gilpin and Waltz profess themselves unable to understand Ashley's important critique.

67 Waltz completely ignores the changing forms of the world economy and the internationalisation of capital, and is therefore blind to the real extent of the changing sifnificance of world trade. By the mid-1980s the value of overseas production by transnational corporations exceeded that of world trade, and much of the latter was intra-firm exchange. The implications of Waltz's failure to consider the forms and extent of the internationalisation of finance and production are far-reaching, but it would be tedious to dwell on this here.

68 Waltz, *Theory of International Politics*, p. 142.

69 Ibid., p. 146.

70 Ibid., p. 163.

71 Ibid., p. 169.

72 See ibid., pp. 205–6.

73 Keohane, *After Hegemony*, pp. 25–30.

74 Waltz, 'Reflections on *Theory of International Politics*', in Keohane (ed.), *Neorealism and its critics*, p. 343.

75 Ibid., p. 344. Keynes had the measure of such thinking when he wrote that: 'The great events of history are often due to secular changes in the growth of population and other fundamental economic causes, which, escaping by their gradual character the notice of contemporary observers, are attributed to the follies of statesmen or the fanaticism of atheists'. Quoted from *The Economic Consequences of the Peace* as an epigraph in Gilpin's *War and Change in World Politics*.

76 John Gerard Ruggie, 'Continuity and transformation in the world polity: Towards a neorealist synthesis', in Keohane (ed.), *Neorealism and its Critics*. Ruggie notes that: 'The problem with Waltz's posture is that, in any social system, structural change itself ultimately has no source other than unit-level processes. By banishing these from the domain of systemic theory, Waltz also exogenizes the ultimate source of systemic change. . . . As a result, Waltz's theory of "society" contains only a reproductive logic, but no transformational logic'. (p. 152).

77 Robert O. Keohane, 'Theory of world politics: structural realism and

beyond', in idem (ed.), *Neorealism and its critics*. Keohane argues that: 'the link between system structure and actor behavior is forged by the rationality assumption', and that 'the most parsimonious version of a structural theory would hold that any international system has a single structure of power. In such a conceptualization, power resources are homogenous and fungible'. (p. 167).

78 Robert O. Keohane and Joseph S. Nye, Jr, *'Power and Interdependence revisited'*, in *International Organization*, 41, 1987, p. 746.

79 Stephen D. Krasner, *Defending the National Interest*, Princeton University Press, Princeton 1978. It is one of the many merits of Krasner's work to have registered the problem that any discussion of hegemony poses for realism; for Waltz's reflections on the management of international affairs, see *Theory of International Politics*, pp. 194–210.

80 For a spirited defense of this procedure, see Hahn, 'On the notion of equilibrium in economics'.

81 Cf. the following comment by Wassily Leontief: 'Page after page of professional economic journals are filled with mathematical formulas leading the reader from sets of more or less plausible but entirely arbitrary assumptions to precisely stated but irrelevant theoretical conclusions', in his foreword to Alfred S. Eichner (ed.), *Why Economics is not yet a Science*, Macmillan, London 1983, viii. Susan Strange has rightly warned against the danger for international relations in seeking to ape the procedures of economics, *'Cave! hic dragones'*, pp. 338–9.

82 See Imre Lakatos, 'Falsification and the methodology of scientific research programmes', in Imre Lakatos and Alan Musgrave (eds), *Criticism and the Growth of Knowledge*, Cambridge University Press, Cambridge 1970; and also the major interpretative essay by Anthony Giddens, 'Positivism and its critics', in idem., *Studies in Social and Political Theory*, Hutchinson, London 1977. (I am of course aware that in the sense of the positivism of the Vienna Circle Popper was not a positivist.).

83 Fred Halliday, 'State and society in international relations: A second agenda', in *Millenium: Journal of International Relations*, 16, 1987, p. 217.

84 The last phrase was coined by Joan Robinson in her discussion of the concept of utility in neoclassical economics, *Economic Philosophy*, Penguin, Harmondsworth 1962, p. 48.

85 Halliday, 'State and society in international relations', p. 218. A definition taken from the recent debate on the state in Marxism and (neo-Weberian) comparative historical sociology. See, for example, Evans et al. (eds), *Bringing the State Back In*.

86 See, of course, Thomas S. Kuhn, *The Structure of Scientific Revolutions*, University of Chicago Press, Chicago 1970. For a superb discussion of these debates, see Richard J. Bernstein, *Beyond Objectivism and Relativism*. Basil Blackwell, Oxford 1983, especially part 2, 'Science, rationality, and incommensurability'; and also Roy Bhaskar, *A Realist Theory of Science*, Harvester, Brighton 1978.

87 See, for example, Richard J. Bernstein, *The Restructuring of Social and Political Theory*, Harcourt Brace Jovanovitch, New York 1976 and Anthony Giddens, *New Rules of Sociological Method*, Hutchinson, London 1976.

88 Bhaskar, *Scientific Realism and Human Emancipation*, pp. 129–36.

89 Karl Popper's argument on this point remains one of the clearest statements: see 'Prediction and prophecy in the social sciences', in *Conjectures and Refutations*, Routledge and Kegan Paul, London 1963, pp. 336–46; see also idem, *The Poverty of Historicism*, Routledge and Kegan Paul, London 1957.

90 Exponents of Critical Theory might suggest that critical knowledge in the service of an emancipatory interest would be a more relevant criterion. However, the fullest working through of this position to date, the work of Habermas, cannot dispense with the need for epistemological criteria beyond those of a purely procedural kind. Thus far even Habermas' most sympathetic critics conclude that his notions of an 'ideal speech situation', the theories of 'universal pragmatics' and of 'communicative action', and his more general philosophical stance have not secured a theory of truth which guards against relativism. See, in particular, Raymond Geuss, *The Idea of a Critical Theory*, Cambridge University Press, Cambridge 1981; and John B. Thompson, 'Universal pragmatics' and 'Rationality and social rationalization', in idem, *Studies in the Theory of Ideology*.

91 Bhaskar, *Scientific Realism and Human Emancipation*, pp. 118–29 and *The Possibility of Naturalism*.

92 Andrew Sayer, *Method in Social Science*, Hutchinson, London 1984, p. 107.

93 Ibid., p. 126. Sayer also makes an important point about assumptions and predictions in the social sciences in this context:

> If predictions and calculations are needed . . . assumptions need not be realistic . . . If explanation is the primary goal, two possibilities exist: 1 If the model is based on rational abstractions and assumptions merely serve to hold constant certain well-defined necessary relations and to 'hold-off' contingent interfering processes, then it may effectively explain (provided it is backed up by qualitative, causal anaysis) some of the constitutive processes in concrete open systems. . . . 2 Alternatively, the assumptions may be 'unrealistic' in the more serious sense that they deny what are known to be necessary (and relevant) features of the system of interest . . . From the point of view of explanation, the effects of relaxing assumptions of either type are very different: in 1 it leaves the characterization of the basic structures and mechanisms represented in the model intact, though their effects at more concrete levels may be modified; in 2, it can leave it in ruins and hence the 'unrealistic' nature of the assumptions is a serious problem as regards their use for illuminating real objects, (p. 170).

94 Cf. Giddens's critical comments on structural sociology and methodological individualism in *The Constitution of Society*, especially pp. 207–13 and 225, f.n. 13. What Giddens writes of Blau's structural sociology applies equally to Waltz: 'Structure, or structural properties, or "structural parameters", exist only in so far as there is continuity in social reproduction across time

and space. And such continuity in turn exists only in and through the reflexively monitored activities of situated actors, having a range of intended and unintended consequences' (p. 212). Or again, in structuration theory, 'structure is implicated in that very "freedom of action" which is treated as a residual and unexplicated category in the various forms of "structural sociology"' (p. 174).

95   Anthony Giddens, 'Functionlism: *apres la lutte*', in idem, *Studies in Social and Political Theory*, Hutchinson, London 1977, p. 116.

96   From the considerable body of I. Wallerstein's work, see *Historical Capitalism*, Verso, London 1983 and *The Politics of the World-Economy*, Cambridge University Press, Cambridge 1984 for accessible and brief overviews.

97   See Walt Rostow, *The Stages of Economic Growth*, Cambridge University Press, Cambridge 1960 and *The World Economy*, University of Texas Press, Austin 1978. Eric Hobsbawm has noted of this perspective that it 'sees the world economy as essentially a collection of "dynamic interacting national economies" which are essentially similar in their objectives . . . [and which] can be either regarded or essentially analysed as a "complex of national economies". . . . It is a world in which the basic problems, which are those of growth, are essentially soluble everywhere by the same methods of economic engineering because they are all equally "economies"', 'The development of the world economy', in *Cambridge Journal of Economics*, 3, 1979, p. 307.

98   John B. Thompson, 'Ideology and Methods of discourse analysis' and 'The theory of structuration', in idem, *Studies in the Theory of Ideology*, Polity Press, Cambridge 1984.

99   Ibid., p. 129.

100   Ibid., pp. 129–30.

101   John Lovering, 'Militarism, capitalism, and the nation-state: towards a realist synthesis', in *Environment and Planning D: Society and Space*, 5, p. 294. See also the valuable exposition and commentary on Giddens in Justin Rosenberg, 'Giddens' 'Nation-State and Violence' in *Millennium: Journal of International Relations*, 19, 1990 (in press).

102   Giddens, *The Nation-State and Violence*, pp. 261–4.

103   Waltz notes that the United States and the USSR differ in the international spread of their economies but inevitably fails to see the significance of this for his understanding of the state; see *Theory of International Politics*, p. 151.

104   In using these terms from Louis Althusser's seminal essay 'Contradiction and overdetermination', in idem, *For Marx*, Verso, London 1979, I do not intend to imply either support for his broad philosophical standpoint or his structuralist borrowings. Rather I want to indicate the importance of Althusser's contribution to the development of a non-reductionist Marxist theory of politics and ideology. For the first study in English to do justice to Althusser's *oeuvre*, see Gregory Elliott, *Althusser – The Detour of*

Theory, Verso, London 1987. See also Giddens, *Central Problems in Social Theory*, pp. 155–60 and Sayer, *Method in Social Science*, pp. 99–100.

105 See Anthony Giddens, 'Time and social organization' in idem, *Social Theory and Modern Sociology*, p. 154. Michael Mann's work is often, and in my view mistakenly, assimilated to that of Giddens. (See his article 'The autonomous power of the state', and also *The Sources of Social Power*, vol. 1, Cambridge University Press, Cambridge 1986.) For although there are marked similarities of approach (see no. 106), Mann does not develop a theory of the international in the way that Giddens attempts and so, like the dualist approach he seeks to supersede, often operates with an implictly realist theory of international politics.

106 Indeed, both Giddens and Mann exhibit a recurrent tendency to slide from distinguishing analytically the administrative and the allocative dimensions of organised power to treating state power (a substantive concept) purely or primarily in terms of its authoritative relations. This can often result in an overestimation of the autonomy of state power as well as an incipient collapsing of the concept of society into that of the state. Mann's formulations on this point are simply ambiguous. On the one hand, he argues that 'the state does not possess a distinctive means of power independent of, and analogous to, economic, military and ideological power. . . . However, the power of the state is irreducible in quite a different socio-spatial and organizational sense. Only the state is inherently centralized over a delimited territory over which it has authoritative power. . . . [And] this in itself does not confer a significant degree of actual power on the state elite'. (See 'The autonomous power of the state', pp. 122–3 and 125.) On the other, Mann speaks of 'territorial centralization' as providing a 'potentially independent basis of power mobilization, being necessary to social development and uniquely in the possession of the state itself' (p. 125). The problem here is the term 'possession': for the authoritative or infrastructural aspects of state power are not something possessed by the state or its elite – in fact, the bulk of Mann's discussion makes this clear, in particular his insistence that despotic and infrastructural power are not enhanced in equal measure. I also remain sceptical of the degree of automony accorded to ideological power by mann in particular. His main historical example, the role played by Christianity, has been subjected to a number of telling objections. See Perry Anderson, 'Those in authority', in *Times Literary Supplement*, 12 December 1986; and Chris Wickham, 'Historical materialism, historical sociology', in *New Left Review*, no. 171, 1988.

107 Anthony Giddens, 'Nine theses on the future of sociology', in idem, *Social Theory and Modern Sociology*, p. 35.

108 Thompson, 'Theories of ideology and methods of discourse analysis', p. 128.

109 Ralph Miliband, *The State in Capitalist Society*, Weidenfeld and Nicolson, London 1969.

110 Nicos Poulantzas, *Political Power and Social Classes*, New Left Books, London 1973.

[111] Nicos Poulantzas, *State, Power, Socialism*, Verso, London 1978; Goran Therborn, *What does the Ruling Class do when it Rules?*, New Left Books, London 1978; John Holloway and Sol Picciotto (eds), *State and Capital*, Edward Arnold, London 1978. For general surveys of Marxist state theory, see Martin Carnoy, *The State and Political Theory*, Princeton University Press, Princeton 1984; and Bob Jessop, *The Capitalist State*, Martin Robertson, Oxford 1982.

[112] See, in particular, Ellen Meiksins Wood, 'The separation of the economic and the political in capitalism', in *New Left Review*, no. 127, 1981, pp. 66–95. Paul Cammack has accurately noted that much of the 'bringing the state back in 'literature

> tend[s] to present 'state' and 'society' as separate polar opposites, denying in practice the presence of classes and class struggle within the state . . . A two-stage process takes place in which first social classes are dissolved into 'society', then this undifferentiated 'society' is counterposed to 'the state'. As a consequence, the idea that the state is differentially penetrated by conflicting classes, and incorporates, reflects and affects the struggle between them, becomes literally unthinkable. (See his 'Review article: Bringing the state back in?', in *British Journal of Political Science*, 19, 1989, pp. 263–4 and 289).

[113] Robert W. Cox, *Production, Power, and World Order*, Columbia University Press, New York 1987. While sympathetic to aspects of Cox's overall approach, I feel that he devotes far too little attention to the relative autonomy of the politics of sovereignty in the modern nation-state system, and he scarcely discusses the world military order. None the less, Cox's work is an important contribution to developing a more adequate theory of international relations.

[114] For an attempt to develop such a model with respect to contemporary British domestic politics, see Bob Jessop, Kevin Bonnett, Simon Bromley and Tom Ling, *Thatcherism*, Polity Press, Cambridge 1988, especially chapter 9.

CHAPTER 2 US HEGEMONY: OIL AS A STRATEGIC COMMODITY

[1] Susan Strange, *States and Markets*, Pinter, London 1988, p. 191.

[2] Raymond Vernon, *Sovereignty at Bay*, Penguin, Harmondsworth 1973, p. 37.

[3] Ibid., pp. 44 and 53–60.

[4] Rhys Jenkins, *Transnational Corporations and Uneven Development*, Methuen, London 1987, p. 21. This book is probably the best Marxist treatment of theories regarding TNCs available.

[5] Vernon, *Sovereignty at Bay*, p. 44.

[6] From a vast literature, see, for example, N. H. Jacoby, *Multinational Oil*, Macmillan, London 1974 and Malcolm Slesser, *Energy in the Economy*, Macmillan, London 1978.

7   M. Adelman, *The World Petroleum Market*, Johns Hopkins Press, Baltimore 1972.

8   Alessandro Roncaglia, *The International Oil Market*, J. A. Kregel (ed.), Macmillan, London 1985, p. 34. Roncaglia's work constitutes a rare and welcome treatment of the oil industry from the Sraffian tradition, one of the most cogent treatments of the industry's economics available.

9   Ibid., p. 147, n. 14.

10 See ibid., pp. 36–7.

11 Paul H. Frankel, *Essentials of Petroleum*, Frank Cass, London 1969.

12 Edith Penrose, *The Growth of Firms, Middle East Oil and Other Essays*, Frank Cass, London 1971.

13 Edith Penrose, 'The development of crisis', in *Daedalus*, 104, 1975, pp. 39–57.

14 Ernest J. Wilson, III, 'World politics and international energy markets', in *International Organization*, 41, 1987, pp. 125–49.

15 Ibid., p. 139.

16 Ibid., p. 143, drawing on Richard Caves, 'Industrial organization, corporate strategy and structure', in *Journal of Economic Literature*, 18, 1980.

17 Ibid., p. 144.

18 Quoted in Jenkins, *Transnational Corporations and Uneven Development*, p. 33.

19 Ibid., pp. 33–6, as well as Ben Fine and Laurence Harris, *Rereading Capital*, Macmillan, London 1978, pp. 146–70.

20 Ibid., pp. 34–5.

21 John Weeks, *Capital and Exploitation*, Edward Arnold, London 1981, p. 153.

22 J. Clifton, 'Competition and the evolution of the capitalist mode of production', in *Cambridge Journal of Economics*, 1, 1977, pp 137–51; and David Harvey, *The Limits to Capital*, Basil Blackwell, Oxford 1982, pp. 204–329.

23 Jenkins, *Transnational Corporations and Uneven Development*, p. 47.

24 Cf. Ben Fine, 'On Marx's theory of agricultural rent', in *Economy and Society*, 8, 1979, pp. 241–78. See also Chibuzo Nwoke, *Third World Minerals and Global Pricing*, Zed Books, London 1987. The Sraffian-inspired critique of the labour theory of value, as in Ian Steedman, *Marx after Sraffa*, Verso, London 19777, thus applies only to the labour-embodied theory of value.

25 Roncaglia, *The International Oil Market*, pp. 42–6.

26 Ibid., . 45.

27 Fine and Harris, *Rereading Capital*, pp. 149–60.

28 See David M. Gordon, Richard Edwards and Michael Reich, *Segmented Work, Divided Workers*, Cambridge University Press, Cambridge 1982, pp. 18–47.

29 For an interesting and important discussion of this and related issues, see Bob Jessop, 'The capitalist state and the rule of capital: problems in the

analysis of business associations', in *West European Politics*, 6, 1983, pp. 139–62.

[30]   P. Nore, 'Oil and contemporary capitalism', in F. Green and P. Nore (eds), *Issues in Political Economy*, Macmillan, London 1979; and 'Oil and the state: a study of nationalization in the oil industry', in P. Nore and T. Turner (eds), *Oil and Class Struggle*, Zed Press, London 1980.

[31]   Nore, 'Oil and contemporary capitalism', p. 113.

[32]   Cyrus Bina, *The Economics of the Oil Crisis*, Merlin Press, London 1985. See also idem, 'Some controversies in the development of rent theory: the nature of oil rent', in *Capital and Class*, no. 39, 1989, pp. 82–112.

[33]   Ibid., p. 77.

[34]   Ben Fine and Laurence Harris, *The Peculiarities of the British Economy*, Lawrence and Wishart, London 1985, pp. 82–9.

[35]   Ibid., pp. 88–9.

[36]   Ibid., p. 87.

[37]   Ibid.

[38]   Stephen D. Krasner, *Defending the National Interest*, Princeton University Press, Princeton 1978. I cannot resist noting that realism has also served as the conduit for one of the most hysterical accounts of recent events in the oil industry. Thus, Hans J. Morgenthau argued that the role of oil as an indispensable raw material for the economy and the military meant that:

> It is now a material factor whose very possession threatens to overturn centuries-old patterns of international politics. . . . [The functional relationships between political, military and economic power] have been disturbed – one might even say destroyed – by the recent use of oil as a political weapon. . . . [The use of the oil weapon] could reduce Japan to the status of satellite, a dependency of the oil-producing nations. . . . . [And so Morgenthau goes on to propose a] crash programme, after the model of the Manhatten Project . . . to develop alternative source of energy. *Politics among Nations* (6th edn with Kenneth W. Thompson), Alfred A. Knopf, New York 1985, pp. 133–5.

Presumably the author saw Nixon's televised address on 7 November 1973:

> Let us unite in committing the resources of this nation to a major new endeavor, an endeavor that in this bicentennial era we can appropriately call 'Project Independence' . . . Let us set as our national goal, in the spirit of Apollo, with the determination of the Manhattan Project, that by the end of this decade we will have developed the potential to meet our own energy needs without depending on any foreign enemy – I mean, energy – sources . . . We have an energy crisis, but there is no crisis of the American spirit. (Quoted with the characteristic and revealing slip excised in Hans Jacob Bullberg, *American International Oil Policy*, Francis Pinter, London 1987, p. 3; for the full text, see Ian Skeet, *OPEC: Twenty-five Years of Prices and Politics*, Cambridge University Press, Cambridge 1988, p. 117.)

[39]   Nicos Poulantzas, *Political Power and Social Classes*, New Left Books, London 1973; see also Harry Magdoff, *The Age of Imperialism*, Monthly Review Press, New York 1969, and James O'Connor, *The Fiscal Crisis of the State*, St Martin's Press, New York 1973.

[40] Krasner, *Defending the National Interest*, p. 148. In view of the criticisms of this book made below it is only fair to point out that Krasner does at one point note that 'it is also necessary to recognize that private actors operated within a larger international economic structure created and sustained by state power. After the Second World War the United States ordered the international economic environment' (p. 348). But Krasner does not appear to recognize the significance of this and makes no attempt to theorize these broader structures. Also important to note is Krasner's liberal assumption that capitalist interests are equivalent to the sanctity of free competition. From a Marxian standpoint the logic of capitalism is characterized by ceaseless accumulation in search of profit. And, as Wallerstein has pointed out, 'capitalists seek profits, maximal profits, in order to accumulate capital . . . They are thereby not merely motivated but structurally forced to seek *monopoly positions*, something which pushes them to seek profit-maximisation via the principal agency that can make it enduringly possible, the *state* (my emphasis), 'The bourgeois(ie) as concept and reality', in *New Left Review*, no. 167, 1988, p. 103. The oil comapnies were and are one of the most successful modern examples of this recurrent tendency to seek to turn profit into rent through the agency of the state.

[41] Ibid., p. 319; see also pp. 346–7.

[42] Ibid., pp. 278 and 334–5.

[43] Ibid., p. 153.

[44] Nicos Mouzelis, 'Marxism or Post-Marxism?', in *New Left Review*, no. 167, 1988, p. 117.

[45] While the contributions to the state-derivation debate avoid the post-Althusserian traps identified by Mouzelis, they run into similar problems at a later stage. The starting point of this school of thought is the argument that, to take a representative formulation from Poulantzas, 'The separation [between the state and the economy] is nothing other than the capitalist form of the presence of the political in the constitution and reproduction of the relations of production' (*State, Power, Socialism*, Verso, London 1978, pp. 18–19). In other words, one must recognize that within the capitalist mode of production there is a 'division of labour in which the two moments of capitalist exploitation – appropriation and coercion – are allocated separately to a 'private' appropriating class and a specialized 'public' coercive institution, the state' (Ellen Meiksins Wood, 'The separation of the economic and the political in capitalism', in *New Left Review*, no. 127, 1981, pp. 81–2). Now, all that this establishes – and it is important given some of the recent and indiscriminate tendencies to autonomise state power – is that the partial institutional separation of the state and the economy under capitalism does not cancel the class character of the state form. (On this, see especially Goran Therborn, *What does the Ruling Class do when it Rules?*, New Left Books, London 1978.) But it does not follow that state power can be reduced to or wholly explained in terms of class power, any more than state power is purely concerned with the mobilization of administrative or authoritative resources.

46 See, for example, Krasner, *Defending the National Interest*.
47 For a detailed and sophisticated disucssion of these issues, which none the less remains firmly within the realist problematic, see Robert O. Keohane, *After Hegemony*, Princeton University Press, Princeton 1984.
48 This model has been most fully explicated by Robert Gilpin, *War and Change in World Politics*, Cambridge University Press, Cambridge 1981.
49 Keohane, *After Hegemony*, p. 32.
50 Ibid., p. 139.
51 Ibid., p. 140.
52 Ibid., p. 141.
53 Ibid., p. 190.
54 Ibid., p. 194.
55 Hans Jacob Bull-Berg, *American International Oil Policy*, Frances Pinter, London 1987, p. 4.
56 Ibid., p. 138.
57 G. John Ikenberry, 'The irony of state strength: comparative responses to the oil shocks in the 1970s', in *International Organization*, 40, 1986, pp. 105–37. For a more detailed treatment of the state in general from this perspective, see John A. Hall and G. John Ikenberry, *The State*, Open University Press, Milton Keynes 1989.
58 Ibid., p. 106.
59 Ibid.
60 Ibid., p. 121.
61 Ibid., p. 136.
62 Ibid.
63 Ibid., p. 111.
64 For a treatment which does just this, but otherwise provides an illuminating account, see Barry Buzan, *An Introduction to Strategic Studies*, Macmillan/ IISS, London 1987.
65 Bruce Russett, 'The mysterious case of vanishing hegemony; or, Is Mark Twain really dead?, in *International Organization*, 39, 1985, pp. 209–10.
66 See K. Oye, 'International Systems Structure and American Foreign Policy', in K. Oye, R. Lieber and D. Rothchild (eds), *Eagle Defiant*, Little, Brown, Boston 1983; and R. Parboni, 'The dollar weapon: From Nixon to Reagan', in *New Left Review*, no. 158, 1986, pp. 5–18.
67 Samuel P. Huntington, 'The U.S. – decline or renewal?', in *Foreign Affairs* 67, 1988–89, p. 84. Huntington is probably in error to suggest that Britain never achieved such predominance, but the circumstances were so radically different as to render the comparison meaningless anyway.
68 Keohane, *After Hegemony* takes the former position, while Robert Gilpin, *The Political Economy International Relations*, adopts the more pessimistic view. See also Susan Strange, *States and Markets*, Pinter, London 1988, pp. 235–40.
69 The following paragraphs on the imperialist epoch draw directly on the magisterial account by Eric J. Hobsbawm, *The Age of Empire, 1875–1914*,

Weidenfeld and Nicolson, London 1987.
70 Eric J. Hobsbawm, *The Age of Capital, 1848–1875*, Weidenfeld and Nicolson, London 1975, p. 160.
71 Hobsbawm, *The Age of Empire, 1875–1914*, pp. 68–9. The significance of cotton can be seen in the fact that: 'In 1880, textiles and clothing were 55.7 per cent of world trade in manufactures by value. In that sector Britain was still in 1880 responsible for 46.3 per cent of world exports: in cotton alone perhaps 80 per cent.' See N. F. R. Crafts, *British Economic Growth during the Industrial Revolution*, Clarendon Press, Oxford 1985, p. 144.
72 Ibid., p. 318 and see pp. 276–9.
73 Ibid., p. 319.
74 For the economic consequences of the First World War, see Gerd Hardach, *The First World War, 1914–1918*, Allen Lane, London 1977.
75 Derek H. Aldcroft, *From Versailles to Wall Street, 1919–1929*, Allen Lane, London 1977, p. 281.
76 Charles P. Kindleberger, *The World in Depression, 1929–1939* (revised edn), University of California Press, Berkeley 1986, pp. 279–80 and 289. This was, of course, the original statement of the theory of hegemonic stability. Kindleberger argued that a single power was needed to stabilize the system, by discharging five functions:

(1) maintaining a relatively open market for distress goods;
(2) providing countercyclical, or at least stable, long-term lending;
(3) policing a relatively stable system of exchange rates;
(4) ensuring the coordination of macroeconomic policies;
(5) acting as a lender of last resort by discounting or otherwise providing liquidity in financial crisis.

77 R. J. Overy, *The Origins of the Second World War*, Longman, London 1987, p. 65. See also Ernest Mandel, *The Meaning of the Second World War*, Verso, London 1986, pp. 11–46.
78 William Carr, *Poland to Pearl Harbor*, Edward Arnold, London 1985, p. 104.
79 Ibid., p. 154.
80 David P. Calleo, *Beyond American Hegemony*, Wheatsheaf, Brighton 1987, p. 132.
81 Geoffrey Barraclough, *An Introduction to Contemporary History*, Penguin, Harmondsworth 1967, pp. 107–8. The events Barraclough refers to included the Boxer Rebellion, the Anglo-Japanese alliance of 1902 and the Russo-Japanese War of 1904–5. John Lewis Gaddis has also noted that the United States's 'assertion of an . . . interest in economic access to and in the territorial integrity of China ran up against the sphere of influence the Russians had constructed through their own arrangements with the Chinese dating from 1896, arrangements which were unilaterally expanded upon four years later in the wake of the Boxer Rebellion'. See 'Legacies: Russian–American relations before the Cold War', in Gaddis, *The Long Peace*, Oxford University Press, Oxford 1987, p. 7.

82 In this context, the First World War had contradictory effects: on the one hand, in Russia a revolutionary movement embarked on a programme of social transformation, sued for peace and instituted a revolutionary foreign policy; and on the other, the uneasy alliance of Zimmerwald split, with the International dividing between a minority Bolshevik faction and a 'patriotic' majority. And with the reconstitution of the Second International by the Right at Berne (February 1919) and the formation of the Third International in Moscow (March 1919), the subsequent divisions among labour parties and union federations were inevitable.

83 Ibid., p. 154.

84 On this, see Victor G. Kiernan, *European Empires From Conquest to Collapse, 1815–1960*, Fontana, London 1982, pp. 208–9.

85 Calleo, *Beyond American Hegemony*, pp. 129–49 is a notable exception to this general consensus. In particular, he argues that the consolidation of states in the Third World, together with the capacity of global capitalism to prosper without territorial empires, render the lessons of interwar hegemonic instability redundant for the present: 'many of the lessons of the past, as seen through hegemonic or Leninist perspectives, seem of questionable relevance'. However, Calleo does not consider the equally important changes wrought by the alteration of the systemic character of world politics. By contrast, Fred Halliday, 'Theorising the international', in *Economy and Society*, 18, 1989, p. 350, criticizes Keohane precisely because he fails to register 'that Britain was dominant in an international system characterised solely by intercapitalist rivalry . . . [whereas] the USA has exerted hegemony over a capitalist world which it claims to protect, as a whole, from the USSR and its allies'.

86 See Giovanni Arrighi, 'A crisis of hegemony', in Samir Amin, Giovanni Arrighi, Andre Gunder Frank and Immanuel Wallerstein, *Dynamics of Global Crisis*, Macmillan, London 1982, p. 93; and idem, *The Geometry of Imperialism* (revised edn), Verso, London 1983. On the asymmetry of the US economy in relation to the world market, see also Mike Davis's argument that US leadership 'was founded not on a rigid pre-eminence in world trade or on hoardings of portfolio investments, but on the maintenance of robust conditions for accumulation within the domestic US economy . . . [as well as] a uniquely asymmmetrical [relation to the world market]' ('The political economy of late imperial America', in *New Left Review*, no. 143, 1984, p. 10).

87 See also Michael Bleaney, *The Rise and Fall of Keynesian Economics*, Mamillan, London 1985; and Philip Armstrong, Andrew Glyn and John Harrison, *Capitalism since World War II*, Fontana, London 1984.

88 Arrighi, 'A crisis of hegemony', p. 58.

89 See Makoto Itoh, 'Marxist theories of crisis' and 'The inflational crisis of world capitalism', in idem, *Value and Crisis*, Pluto Press, London 1980; but note also the important critique by Simon Clarke, 'The basic theory of capitalism: A critical review of Itoh and the Uno school', in *Capital and*

Class, no. 37, 1989, pp. 133–49. An on the anti-imperialist successes in the periphery during the 1970s, see Fred Halliday, *The Making of the Second Cold War*, Verso, London 1983, pp. 81–104 and idem, *Cold War, Third World*, Century Hutchinson, London 1989, pp. 24–51.

90 Nicos Poulantzas, *Classes in Contemporary Capitalism*, New Left Books, London 1975, p. 87. I do not share Poulantzas's view that the crisis of imperialism betokens revolutionary crises in the core (it was of course quite otherwise in the periphery) and neither do I believe that 'world class struggles' are unified to any significant degree. Rather, I endorse Poulantzas's central point: namely, the current crisis is, analytically speaking, one of capital accumulation on a global scale. For further discussion of this point, see chapter 4.

91 Arrighi, 'A crisis of hegemony', p. 66.

92 See also the penetrating analysis of Susan Strange in her *Casino Capitalism*, Basil Blackwell, Oxford 1986. In this context it is also worth noting that claims for the overextension of the United States are somewhat misplaced. This has been pithily explained by Huntington:

> a rough prototypical allocation of GNP for the three largest economies might be as follows:
>
> – U.S. consumption (private and public) at about 78 percent of its GNP, Japan's at 67 percent, and the Soviety Union's 56 percent;
> – U.S. defense spending at about seven percent of its GNP, Japan's about one percent, and the Soviet Union's 18 percent;
> – U.S. investment at roughly 17 percent of GNP, Japan's at about 30 percent of its GNP, and the Soviet's about 26 per cent.
> In short, the Soviets arm, the United States consumes, Japan invests. . . . If the United States falters economically, it will not be because U.S. soldiers, sailors and airmen stand guard on the Elbe, the Strait of Hormuz and the 38th parallel; it will be because U.S. men, women and children overindulge themselves in the comforts of the good life. Consumerism, not militarism, is the threat to American strength. ('The U.S. – decline or renewal?', pp. 86–8.)

In addition, Francis M. Bator notes that: 'The reduction in net national investment share, from its average value of nine percent during 1950–79 to 2.2 percent in 1987, has been masked to a degree by the fact that two-thirds of the 6.8-percentage-point decline occurred not in net private investment at home, but rather in net *government* civilian investment, and in net U.S. investment *offshore*, a consequence mainly of the shift from net exports to net imports' ('Must we retrench?', in *Foreign Affairs*, 68, 1989, p. 99, n. 10). In turn, the primary cause of the fall in investment, together with the adverse move in net imports associated with the rise of the dollar, was the high real interest rates of Reaganomics. For further discussion of this, see chapter 6.

93 See Buzan, *An Introduction to Strategic Studies*, pp. 291–6.

94 Richard K. Betts, 'Nuclear weapons', in Joseph S. Nye, Jr (ed.), *The*

Making of America's Soviet Policy, Yale University Press, New Haven 1984, p. 98.

95   Donald MacKenzie, 'Nuclear war planning and strategies of coercion', in *New Left Review*, no. 148; 1984, p. 52.

96   Henry Kissinger, *Years of Upheaval*, Michael Joseph, London 1982, quoted in Greville Rumble, *The Politics of Nuclear Defence*, Polity Press, Cambridge 1985, p. 117.

97   Thus, Buzan, *An Introduction to Strategic Studies*, pp. 168–94, argues that the assault on MAD came from two directions: first, some argued for an extension of the logic of limited nuclear war to a full-blown doctrine of extended deterrence through the maintenance of escalation dominance; and second, others argued for a complementary escape from MAD through the acquisition of strategic defences.

CHAPTER 3 FROM BRITISH TO UNITED STATES LEADERSHIP

1   For a more detailed exposition, see Alessandro Roncaglia, *The International Oil Market*, Macmillan, London 1985, pp. 8–22.

2   David Harvey, *The Limits to Capital*, Basil Blackwell, Oxford 1982, pp. 330–72 contains a valuable discussion of the general case here. Other important discussions of the economics of oil include M. Massarrat, 'The "Energy crisis": The struggle to redistribute surplus profit' and P. Nore, 'Oil and the state: A study of nationalization in the oil industry' in P. Nore and T. Turner (eds), *Oil and Class Struggle*, Zed press, London 1980; and P. Nore, 'Oil and Contemporary Capitalism', in F. Green and P. Nore (eds), *Issues in Political Economy*, Macmillan, London 1979.

3   Roncaglia, *The International Oil Market*, p. 45.

4   H. Williamson, R. Andreano, A. Daum and G. Klose, *The American Petroleum Industry: The age of illumination, 1859–1899*, Northwestern University Press, Evanston 1959, p. 728.

5   H. Williamson et al., *The American Petroleum Industry: The age of energy, 1899–1959*, Northwestern University Press, Evanston 1963, pp. 6–7.

6   Edward Shaffer, *The United States and the Control of World Oil*, Croom Helm, London 1983, pp 23–37.

7   Williamson et al., *The age of illumination*, p. 729.

8   Williamson et al., *The age of energy*, p. 7.

9   See David S. Landes, *The Unbound Prometheus*, Cambridge University Press, Cambridge 1972, p. 281 and also Williamson et al., *The age of illumination*, pp. 725–6 and 730.

10   Williamson et al., *The age of energy*, p. 14; and see also John Blair, *The Control of Oil*, Pantheon Books, New York 1976, p. 127.

11   For an analysis along these lines, see Michael Tanzer, *The Race for Resources*, Monthly Review Press, New York 1980.

12   On the significance of this development, see Kees van der Pijl, *The Making of an Atlantic Ruling Class*, Verso, London 1984.

13 See, in particular, John Agnew, *The United States in the World Economy*, Cambridge University Press, Cambridge 1987, pp. 58–63.
14 Shaffer, *The United States and the Control of World Oil*, p. 35.
15 See the pioneering analysis of Venn, *Oil Diplomacy in the Twentieth Century*, pp. 14–34.
16 Ibid., p. 39.
17 Quoted in ibid., p. 49.
18 For details, see Shaffer, *The United States and the Control of World Oil*, pp. 45–70.
19 William Carr, *From Poland to Pearl Harbor*, Edward Arnold, London 1985, pp. 1–2.
20 David S. Painter, *Private Power and Public Policy*, I. B. Tauris, London 1986, p. 4. See also Louis Turner who notes that: 'From the start of the serious search for oil, there was growing involvement of the parent governments, which was stimulated by the 1914–18 war and culminated in heavy postwar diplomatic infighting around oil rights, particularly in the Middle East', *Oil Companies in the International System*, RIIA/George Allen & Unwin, London 1983, p. 22.
21 Venn, *Oil Diplomacy in the Twentieth Century*, p. 74. See also chapter 5.
22 See Blair, *The Control of Oil*, p. 32.
23 See R. Eden, M. Posner, R. Bending, E. Crouch and J. Stanislaw, *Energy Economics*, Cambridge University Press, Cambridge 1981, p. 74.
24 For further details, see G. Luciani, *The Oil Companies and the Arab World*, Croom Helm, London 1984; and Roncaglia, *The International Oil Market*.
25 Eden et al., *Energy Economics*, p. 247.
26 F. Al-Chalabi, *OPEC and the International Oil Industry*, Oxford University Press, Oxford 1980, p. 13.
27 Blair, *The Control of Oil*, p. 57.
28 Roncaglia, *The International Oil Market*, p. 55.
29 Blair, *The Control of Oil*, p. 52.
30 Ibid., p. 128.
31 Ibid., p. 189.
32 On the differences between monopoly and state monopoly capitalism, see B. Fine and L. Harris, *Rereading Capital*, Macmillan, London 1979, pp. 114–45.
33 B. Fine and L. Harris, *The Peculiarities of the British Economy*, Lawrence and Wishart, London 1985, p. 84.
34 See the valuable analysis of Nicoline Kokxhoorn, *Oil and Politics*, Lang, Frankfurt/M. 1977.
35 Eden et al., *Energy Economics*, p. 77. Overall, the US position was more favourable: US companies controlled nearly 40% of overseas production in 1939 and held nearly one half of foreign reserves; see Painter, *Private Power and Public Policy*, p. 14.
36 G. Foley, *The Energy Question*, Penguin, Harmondsworth 1981, p. 67.
37 See N. Chomsky, 'Foreign policy and the intelligentsia', in idem, *Towards*

the *New Cold War*, Sinclair Browne, London 1982; and L. H. Shoup and W. Minter, 'Shaping a new world order: The Council on Foreign Relations' blueprint for world hegemony', in Holly Sklar (ed.), *Trilateralism*, South End Press, Boston 1980.

38   For relevant studies of Soviet foreign policy during this period, see Vojtech Mastny, *Russia's Road to the Cold War*, Columbia University Press, New York 1979; Joseph L. Nogee and Robert H. Donaldson, *Soviet Foreign Policy since World War II*, Pergamon Press, Oxford 1988; and Adam B. Ulam, *Expansion and Coexistence*, Holt, Rinehart and Winston, New York 1974.

39   Studies of US foreign policy drawn upon here include Stephen E. Ambrose, *Rise to Globalism*, Penguin, Harmondsworth 1983; John Lewis Gaddis, *The United States and the Origins of the Cold War, 1941–1947*, Columbia University Press, New York 1972 and *Strategies of Containment*, Oxford University Press, Oxford 1982.

40   Martin McCauley, *The Origins of the Cold War*, Longman, London 1983, p. 49.

41   Quoted in ibid.

42   For a full discussion of these problems, see Alan S. Milward, *The Reconstruction of Western Europe*, Oxford University Press, Oxford 1984.

43   For different interpretations of the 'meaning' of the Cold War, see the realist account of A. DePorte, *Europe between the Superpowers*, Yale University Press, London 1979; the argument of M. Cox which stresses the functional aspects of the Cold War for intrabloc management in 'Western Capitalism and the Cold War System', in M. Shaw (ed.), *War, State and Society*, Macmillan, London 1984, 'The rise and fall of the "Soviet Threat"', in *Politial Studies*, 32, pp. 484–98 1985, and 'The Cold War as a system', in *Critique*, pp. 17–82 1986; and the interpretation of Fred Halliday which stresses the antagonistic rivalry between capitalism and communism as global systems albeit refracted through the Great Power competition of the United States and the Soviet Union in *The Making of the Second Cold War*, Verso, London 1983.

44   See Cox, 'The Cold War as a system'.

45   Cox, 'The Rise and Fall of the 'Soviet Threat'', p. 48.

46   See, in particular, Eric Hobsbawm, 'The development of the world economy', in *Cambridge Journal of Economics*, 3, 1979, pp. 305–18; and also Michel Aglietta, 'World capitalism in the eighties', in *New Left Review*, no. 136, 1982, pp. 5–41; E. A. Brett, *The World Economy since the War*, Macmillan, London 1985; Stuart Corbridge, Capitalist World Development, Macmillan, London 1986; and Alain Lipietz, 'Towards global Fordism? Marx or Rostow', in *New Left Review*, no. 132, 1982, pp 33–47; and 'How monetarism choked Third World industralization', in *New Left Review*, no. 145, 1984, pp. 71–87.

47   Milward, *The Reconstruction of Western Europe*, xv.

48   For a detailed treatment of these issues, see Philip Armstrong, Andrew

Glyn and John Harrison, *Capitalism since World War II*, Fontana, London 1984; Michael Bleaney, *The Rise and Fall of Keynesian Economics*, Macmillan, London 1985, pp. 92–130; John Eatwell, *Whatever Happened to Britain?*, Duckworth, London 1982, pp. 86–104; and Angus Maddison, *Phases of Capitalist Development*, Oxford University Press, Oxford 1982, pp. 126–36.

49   Robert O. Keohane, 'The world political economy and the crisis of embedded liberalism', in J. Goldthorpe (ed.), *Order and Conflict in Contemporary Capitalism*, Oxford University Press, Oxford 1984, pp. 21 and 30; see also Bob Rowthorn, 'Late capitalism', in idem., *Class, Conflict and Inflation*, Lawrence and Wishart, London 1980.

50   Bleaney, *The Rise and Fall of Keynesian Economics*, p. 115; and on this pattern of growth more generally, see J. Cornwall, *Modern Capitalism*, Martin Robertson, Oxford 1977, pp. 122–36 and 159–94.

51   Aglietta, 'World capitalism in the eighties', p. 18.

52   For a discussion of this point, see Makoto Itoh 'The World Economic Crisis', in *new Left Review*, no. 138, 1983, pp 93–5.

53   See Brett, *The World Economy since the War*, pp. 80–102 and N. Hood and S. Young, *The Economics of Multinational Enterprise*, Longman, London 1979, pp. 10–41.

54   Andrew Carew, *Labour Under the Marshall Plan*, Manchester University Press, Manchester 1988, quoted from Bill Schwarz's review in *Capital and Class*, no. 36, 1988, p. 165.

55   Simon Clarke, *Keynesianism, Monetarism and the Crisis of the State*, Edward Elgar, Aldershot 1988, p. 256, n.2.

56   See A. Kleinknecht, 'Prosperity, crisis and innovation patterns', in *Cambridge Journal of Economics*, 8, 1984, pp. 251–70; and for a fuller discussion, see Venn, *Oil Diplomacy in the Twentieth Century*, pp. 4–6; and P. Odell, 'Draining the world of energy', in R. Johnston and P. Taylor (eds), *A world in Crisis?*, Basil Blackwell, Oxford 1986.

57   Venn, *Oil Diplomacy in the Twentieth Century*, p. 3.

58   Ibid., p. 9. This is the central argument of Venn's important study which none the less fails both to theorize the strategic role of oil and, by focusing on state-to-state and state-to-company relations, to account for the economic and political dynamics of the industry. Venn tends to allow for state policy to be shaped *either* instrumentally by company pressure *or* by broader strategic considerations, or by a mix of the two. But this excludes from view a consideration of the broader relations between economic growth, oil and US hegemony.

59   Ernest Mandel, *The Meaning of the Second World War*, Verso, London 1986, p. 72.

60   Venn, *Oil Diplomacy in the Twentieth Century*, p. 190, n. 1.

61   Robert O. Keohane, *After Hegemony*, Princeton University Press, Princeton 1984, p. 140.

62   Venn, *Oil Diplomacy in the Twentieth Century*, pp. 108–9.

63 Roger Owen, *The Middle East in the World Economy, 1800–1914*, Methuen, London 1981, p. 287.

64 Ibid., pp. 290 and 291.

65 Eric J. Hobsbawm, *The Age of Revolution, Europe 1789–1848*, Weidenfeld and Nicolson, London 1962, p. 133.

66 See Rosemarie Said Zahlan, *The Making of the Modern Gulf States*, Unwin Hyman, London 1989, chapter 2; and Fred Halliday, *Arabia without Sultans* (with Postscript), Penguin, Harmondsworth 1979, pp. 427–31.

67 Zahlan, p. 13.

68 Ibid., p. 10.

69 Dilip Hiro, *Islamic Fundamentalism*, Paladin, London 1988, chapter 5; and Halliday, *Arabia without Sultans*, pp. 47–9.

70 Halliday, *Arabia without Sultans*, p. 467.

71 E. H. Carr, *The Bolshevik Revolution, 1917–1923*, vol. 3, Penguin, Harmondsworth 1966, p. 244.

72 Ibid., p. 465.

73 Ibid., p. 237.

74 Ibid., p. 470.

75 William Stivers, *Supremacy and Oil*, Cornell University Press, Ithaca 1982, p. 76.

76 Ibid., p. 105. The role of air power in this overall strategy should be underscored. Victor G. Kiernan, *European Empires from Conquest to Collapse, 1815–1960*, Fontana, London 1982 has explained this as follows:

> Whatever inadequacies might come to light, air power was establishing itself as the grand new asset. It reflected and accentuated the widening estrangement between rulers and ruled brought about by nationalism and repression. In 1932 at the Disarmament Conference the British government joined the cry against bombardment of towns from the air, but it could not bring itself to renounce its new-found celestial empire. It would not give up 'the use of such machines as are necessary for police purposes in outlying places'. Its representative congratulated himself on having saved the bombing aeroplane. As Lloyd George put it in homely phrase, other countries would have agreed to a ban, 'but we insisted on reserving the right to bomb niggers!'. (p. 200)

77 Stivers, p. 141.

78 Ibid., pp 171–2.

79 See the discussion in ibid., pp. 178–9.

80 Ibid., pp. 68–9 and more generally pp. 49–74. This remained the case through to the postwar period, as later events in Iran were to demonstrate.

81 See Zahlan, *The Making of the Modern Gulf States*, p. 22; and Halliday, *Arabia without Sultans*, pp. 49–50.

82 For a clear treatment of this issue, see Svante Karlsson, *Oil and the World Order*, Berg, Leamington Spa 1986, pp. 60–5. See also Painter, *Private Power and Public Policy*, chapters 3 and 4.

83 State Department official memorandum, quoted in Keohane, *After Hegemony*, p. 153, n. 3.

84 For these events, see Gaddis, *The United States and the Origins of the Cold*

War, pp. 309–12; Nogee and Donaldson, *Soviet Foreign Policy since World War II*, pp. 82–3; and Painter, *Private Power and Public Policy*, pp. 75–81.

85   For a careful discussion of this, see Karlsson, *Oil and the World Order*, pp. 95–104; and on the coup in Iran pp. 119–37, as well as Keohane, *After Hegemony*, pp. 167–9, Stephen D. Krasner, *Defending the National Interest*, Princeton University Press, Princeton 1978, pp. 119–28, and Painter, *Private Power and Public Policy*, chapter 8.

87   Charles A. Kupchan, *The Persian Gulf and the West*, Allen and Unwin, London 1987, p. 15. See Painter, *Private Power and Public Policy*, pp. 116–27 for details.

87   N. Chomsky, *The Fateful Triangle*, Pluto Press, London 1983, p. 20.

88   Ibid.

89   At this time, Eden faced two major challenges abroad: how to respond to developments in continental Europe and how to cope with Nasser in Egypt. Underlying this dilemma was the desire to maintain 'Great Power' status which involved 'the commonwealth, the sterling area, and assured access (via the Suez Canal) to cheap oil paid for in sterling'. The Permanent Under-Secretary at the Foreign Office, Sir Ivone Kirkpatrick, sent a personal note 'to the British Ambassador in Washington after Eisenhower had written to Eden advising him not to make Nasser out to be a bigger man than he was'. He wrote:

> If we sit back while Nasser consolidates his position and graudally acquires control of the oil-bearing countries he can and is, according to our information, out to wreck us. If Middle East oil is denied to us for a year or two, our gold reserves will disappear. If our gold reserves disappear, the sterling area disintegrates. If the sterling area disintegrates and we have no reserves, we shall not be able to maintain a force in Germany or indeed anywhere else. I doubt whether we shall be able to pay for the bare minimum necessary for our defence. And a country that cannot provide for its defence is finished.

See, Keith Kyle, 'Protocols of Sevres' (reviewing Richard Lamb's *The Failure of the Eden Government*, Sidgwick, London 1987), in *London Review of Books*, 21 January 1988.

90   Kupchan, *The Persian Gulf and the West* notes of this, 'That the U.S. presence was more than symbolic was made certain by the placement of tactical air units in Turkey and the movement of a seaborne Marine combat team from Okinawa to the Persian Gulf. The size of the unit in Beirut was also determined by contingency plans to invade Iraq if Baghdad made threatening moves toward Jordan, Kuwait, or Saudi Arabia' (p. 23).

91   See ibid., pp. 25–31.

92   Halliday, *Arabia without Sultans*, p. 71.

93   ibid., p. 74.

94   Michael McGwire, 'The Middle East and Soviet Military Strategy', in *MERIP Reports*, 18, 1988, p. 12. According to McGwire, by the 1970s the Soviets believed that their strategic parity would deter a direct preemptive strike on the Soviet Union and that the military priority was to

deny the United States 'a bridgehead in Europe from which to mount a second phase offensive . . . [and this required] secure sea lanes between eastern and western Soviet territory, and the ability to repulse a Western military buildup not just in Europe but anywhere on the Eurasian continent'. It was this strategic logic which gave North Africa and the Horn its importance.

95 Robin Edmonds, *Soviet Foreign Policy 1962–1973*, Oxford University Press, Oxford 1975, p. 121.

96 Kupchan, *The Persian Gulf and the West*, p. 50.

97 Shaffer, *The United States and the Control of World Oil*, pp. 95–7.

98 Karlsson, *Oil and the World Order*, pp. 90–1.

99 For detailed statistical material, see Eden et al., *Energy Economics*; and Foley, *The Energy Question*. The coal to oil ratios were as follows: in 1937 the figure for the United Kingdom was 2.8:1, for Germany 48.5, for France 11.1 and for Italy 8.8; the respective figures for 1972 were 0.87, 0.67, 0.32 and 0.09.

100 For a detailed examination of this issue, see Blair, *The Control of Oil*, pp. 98–120.

101 Ibid., p. 50; and Roncaglia, *The International Oil Market*, p. 152, n. 21.

102 Ibid., pp. 196–203; and for further discussion, see Roncaglia, *The International Oil Market*, pp. 101–4 and Karlsson, *Oil and the World Order*, pp. 109–13.

103 D. Gisselquist, *Oil Prices and Trade Deficits*, quoted in Fine and Harris, *The Peculiarities of the British Economy*, p. 84.

104 Shaffer, *The United States and the Control of World Oil*, pp. 121–3.

105 On this episode, see Karlsson, *Oil and the World Order*, pp. 158–71: Keohane, *After Hegemony*, pp. 174–7; Richardo Parboni, *The Dollar and its Rivals*, Verso, London 1981, pp. 51–3; and Shaffer, *The United States and the Control of World Oil*, pp. 128–40.

106 See Shaffer, *The United States and the Control of World Oil*, pp. 121–3.

107 See Roncaglia, *The International Oil Market*, pp. 98–9.

108 Venn, *Oil Diplomacy in the Twentieth Century*, p. 135.

109 See Philip Hanson, *Western Economic Statecraft in East–West Relations*, RIIA/Routledge and Kegan Paul, London 1988, pp. 28–9.

110 P. Odell, *Oil and World Power*, Penguin, Harmondsworth 1983, pp. 138 and 216.

111 Data in Parboni, *The Dollar and its Rivals*, p. 53 and IEA/OECD, *Energy Policies and Programmes of IEA Countries*, Paris 1985, pp. 124, 126 and 130.

112 See Al-Chalabi, *OPEC and the International Oil Industry*, p. 112.

113 Ibid., p. 65–6.

114 Ibid., p. 71.

115 Painter, *Private Power and Public Policy*, p. 56.

116 Ibid., p. 9.

117 Ibid., p. 96. At this point, I should note that while Painter's excellent

study of multinational oil companies and US foreign policy between 1941 and 1954 provides the fullest account yet of company-state manoeuvres, it can be criticized for its corporatist focus, for it shares some of the generic weaknesses outlined in chapter 2. Specifically, Painter attributes the impetus to the overseas expansion of the US state interests too exclusively to the desire to conserve the reserves of the Western Hemisphere. What is obscured by this optic is the keen appreciation of the role that Middle East oil would have to play in a renewed world order, thereby not only securing US national interests but also underpinning US hegemony within the global system.

118 The basis for these remarks is the formal Kaleckian model of monopoly, as amended to account for the full effects of capitalist competition. In the Kaleckian model, the degree of monopoly is determined by the level of collusion, the degree of concentration and the price elasticity of demand (see Keith Cowling, *Monopoly Capitalism*, Macmillan, London 1982). However, there are various shortcomings with this approach (see Ben Fine and Andy Murfin, *Macroeconomics and Monopoly Capitalism*, Harvester Press, Brighton 1984, especially part II). On the supply side, oligopoly is seen as an equilibrium divergence from perfect competition in that the degree of monopoly is defined as the difference between price and marginal cost, expressed as a percentage of price. The effect of this is to ignore the role of forces that would tend to move the system away from such an equilibrium, namely changes in cost functions or the redistribution of market shares. In addition, the distribution of costs itself is left unexplained. On the demand side, the categories of wage and profit replace the Keynesian notions of consumption and investment demand, while the level of demand, output and employment is determined by capitalist expenditure from profits. By aggregating from sectors to the whole economy, however, the coercion to invest in competition for reduced costs and increased market shares is neglected, as is the role of the credit system in facilitating the intersectoral mobility of capital.

CHAPTER 4 THE ECONOMIC CRISIS, OIL AND US–OPEC RELATIONS

1 Robert O. Keohane, *After Hegemony*, Princeton University Press, Princeton 1984, p. 190. Keohane, however, clearly recognizes that OPEC has not operated as a cartel.
2 Stephen D. Krasner, *Structural Conflict*, University of California Press, Berkeley 1985, p. 71.
3 Hans Jacob Bull-Berg, *American International Oil Policy*, Frances Pinter, London 1987, p. 4.
4 Hans, J. Morgenthau, *Politics among Nations* (6th edn with Kenneth W. Thompson), Alfred A. Knopf, New York 1985, p. 133.
5 Robert Gilpin, *War and Change in World Politics*, Cambridge University Press, Cambridge 1981, p. 208. Gilpin does at least rightly note that the

impact of this was limited by the fact that Saudi Arabia and Iran were allies of the United States.

6   Norman S. Fieleke, *The International Economy under Stress*, Ballinger, Cambridge, Mass., p. 2.

7   Daniel Yergin and Martin Hillenbrand (eds), *Global Insecurity*, Penguin, Harmondsworth 1983, p. 3.

8   Quoted in Pierre Terzian, *OPEC: The Inside Story*, Zed Press, London 1985, p. 201.

9   Edward Said, *Covering Islam*, Routledge and Kegan Paul, London, 1981.

10 The title by which the OECD report *Towards Full Employment and Price Stability*, Paris 1977 was popularly known.

11 Ibid., p. 51.

12 For a relevant discussion, see David Currie, 'World capitalism in recession', in S. Hall and M. Jacques (eds), *The Politics of Thatcherism*, Lawrence and Wishart, London 1983.

13 For a careful treatment of this, see John Cornwall, *The Conditions for Economic Recovery*, Martin Robertson, Oxford 1983, pp. 162–72.

14 For a theoretical model of this process, see Bob Rowthorn, 'Conflict, inflation and money', in idem, *Capitalism, Conflict and Inflation*, Lawrence and Wishart, London 1980.

15 Michael Bleaney, *The Rise and Fall of Keynesian Economics*, Macmillan, London 1985, p. 173.

16 This is a theme stressed in the important article by Nigel Thrift, 'The geography of international economic disorder', in R. Johnston and P. Taylor (eds), *A World in Crisis?*, Basil Blackwell, Oxford 1986.

17 Makoto Itoh, 'The inflational crisis of world capitalism', in idem, *Value and Crisis*, Pluto Press, London 1980, pp. 156–7.

18 Brett, *International Money and Capitalist Crisis*, Heinemann, London 1983, p. 212; see also Alain Lipietz, *The Enchanted World*, Verso, London 1985.

19 For a discussion of these developments, see Ankie Hoogvelt, 'The new international division of labour', in Ray Bush, Gordon Johnston and David Coates (eds), *The World Order*, Polity Press, Cambridge 1987.

20 See Sam Aaronovitch and Ron Smith (with Jean Gardiner and Roger Moore), *The Political Economy of British Capitalism*, McGraw-Hill, London 1981, pp. 180–1; and for the US economy, see Manuel Castells, *The Economic Crisis and American Society*, Basil Blackwell, Oxford 1980, pp. 78–137: as well as the important debate carried in the *Cambridge Journal of Economics* among Thomas E. Weisskopf, 'Marxian crisis theory and the rate of profit in the postwar U.S. economy', 3, 1979, pp. 341–78; Fred Moseley, 'The rate of surplus value in the postwar US economy: a critique of Weisskopf's estimates' and Weisskopf, 'The rate of surplus value in the postwar US economy: a response to Moseley's critique', 9, 1985, pp 57–79 and 81–4; Moseley, 'The profit share and the rate of surplus value in the US economy, 1975–85', Andrew Henley, 'Labour's shares and profitability crisis in the US: recent experience and post-war trends', and G. Dumenil,

M. Glick and J. Rangel, 'The rate of profit in the United States', II, 1987, 393–8 and 315–30 and 331–59.

21 See Alain Lipietz, 'Towards global Fordism? Marx or Rostow', in *New Left Review*, no. 132, 1982, pp. 36–7 and Lester Thurow, 'America, Europe and Japan', in *The Economist*, 9 November 1985.

22 I. Wallerstein, *The Politics of the World-Economy*, Cambridge University Press, Cambridge 1984, pp. 63–4.

23 The following analysis draws on the provocative work of Alain Lipietz, 'How monetarism choked Third World industrialization', in *New Left Review*, no. 145, 1984, pp. 71–87; and idem, *Mirages and Miracles*, Verso, London 1987.

24 See the careful review of much of the relevant evidence in David Gordon, 'The global economy: New edifice or crumbling foundations?', in *New Left Review*, no. 168, 1988, pp. 24–64.

25 Ibid., p. 57.

26 See the analyses of Calleo, *The Imperious Economy*, Harvard University Press, Cambridge, Mass. 1982 and of Howard M. Wachtel, 'The politics of international money', Transnational Institute 1987.

27 Lipietz, 'How monetarism choked Third World industrialization', p. 82. The figures above come from this article.

28 See W. Andreff, 'The international centralization of capital and the re-ordering of world capitalism', in *Capital and Class*, no. 22, 1984, pp. 63–4.

29 Bleaney, *The Rise and Fall of Keynesian Economics*, pp. 158–60.

30 Ibid., p. 193. This was true in general until the mid-1980s at least.

31 This analysis draws on that of Lipietz, 'Towards global Fordism?', pp. 34–5.

32 The pioneering work here was Michel Aglietta's *A Theory of Capitalist Regulation*, New Left Books, London 1979.

33 James O'Connor, *Accumulation Crisis*, Basil Blackwell, Oxford 1984.

34 Simon Clarke, 'Overaccumulation, class struggle and the regulation approach', in *Capital and Class*, no. 36, 1988, p. 89, n. 1. Clarke's theorization of the overaccumulation of capital is an important contribution, but unless this is connected to developments in the monetary sphere it is hard to explain how routine overaccumulation coheres into a generalized economic crisis.

35 See Thrift, 'The geography of international economic disorder'.

36 Fred Block, 'Political choice and the multiple "logics" of capital', in *Theory and Society*, 15, 1986, pp. 175–92.

37 This contradiction is at the heart of current US dilemmas *vis-à-vis* Japan: on the one hand, the United States requires an inflow of Japanese capital to fund the deficits yet is losing out to Japanese competition in world markets, while on the other, Japan wants to maintain its access to the US market. Of course, the problem is also structured by the bilateral security relationship, with US opinion divided as to whether to trade tutelage for a Japanese military buildup which might ameliorate economic tensions.

Business Week (7 August 1989) reflected this tension in a notably sympathetic treatment of the 'revisionists', entitled 'Rethinking Japan', which noted a marked shift of elite opinion towards those who favour managed trade and increasing bilateralism with Japan. The controversy surrounding the FSX fighter deal is regarded as a turning point: 'For the first time (sic), policymakers treated America's economic strength as a national-security issue: Washington extracted from Tokyo a guarantee of how much of the FSX fighter business the U.S. would get, how much U.S. technology would be shared, and what new technology would flow back.' (Indeed, the FSX deal produced an interesting reversal of bureaucratic alignments in Washington: the Pentagon supported the plan in order to encourage greater Japanese defence effort; the Commerce Department opposed it, fearing Japanese competition in a technology where the United States currently maintains a lead.) But in the corresponding editorial – 'Rethinking Japan: Yes, but' – attention was drawn to the degree of special pleading involved in calls for managed trade alongside warnings of the dangers involved in diverting attention away from US problems, drifting towards protectionism and damaging the alliance.

[38] E. A. Brett, *The World Economy since the War*, Macmillan, London 1985, pp. 113–14. In general, the Triffin paradox states that a given currency cannot simultaneously serve as a means of international payment and a reserve for central banks.

[39] Ibid., p. 118.

[40] For a discussion of the importance of this, see Kees van der Pijl, *The Making of an Atlantic Ruling Class*, Verso, London 1984, p. 258.

[41] Riccardo Parboni, *The Dollar and Its Rivals*, Verso, London 1981, p. 89; see also E. A. Brett, *International Money and Capitalist Crisis*, and Susan Strange, *Casino Capitalism*, Basil Blackwell, Oxford 1986.

[42] David P. Calleo, *The Imperious Economy*, pp. 130–8.

[43] Quoted in Noam Chomsky, 'The cornerstone of the American system', foreword to Yann Fitt, Alexander Faire and Jean-Pierre Vigier, *The World Economic Crisis*, Zed Press, London 1980, p. 1.

[44] See Brett, *The World Economy since the War*, pp. 128–9.

[45] For a fuller discussion of these developments, see Van der Pijl, *The Making of an Atlantic Ruling Class*, pp. 178–213.

[46] Here I draw on the analysis of Michael Cox, 'The Cold War as a system', in *Critique*, no. 17, 1986, pp. 17–82; see also Jonathan Steele, *The Limits of Soviet Power*, Penguin, Harmondsworth 1985, pp. 87–115.

[47] See I. Wallerstein, *The Politics of the World-Economy*, p. 60.

[48] Cyrus Bina, *The Economics of the Oil Crisis*, Merlin Press, London 1985, p. 3.

[49] A. Roncaglia, *The International Oil Market*, Macmillan, London 1985, pp. 83 and 101–4.

[50] Quoted in ibid., p. 102.

51 John Blair, *The Control of Oil*, Pantheon Books, New York 1976, pp. 294–320.
52 Andreff, 'The international centralization of capital and the re-ordering of world capitalism', p. 73; and P. Nore, 'Oil and the state: A study of nationalization in the oil industry', in P. Nore and T. Turner (eds), *Oil and Class Struggle*, Zed Press, London 1980, p. 72.
53 Fiona Venn, *Oil Diplomacy in the Twentieth Century*, Macmillan, London 1986, p. 179, n. 16.
54 Nicoline Kokxhoorn, *Oil and Politics*, Lang, Frankfurt/M. 1977, p. 183; and N. Hood and S. Young, *The Economics of Multinational Enterprise*, Longman, London 1979, p. 40.
55 See in particular P. Odell and K. Rosing, *The Future of Oil*, Kogan Page, London 1983, pp. 83–105.
56 Nore, 'Oil and the state', p. 73.
57 On this, see Terzian, *OPEC: The Inside Story*, pp. 188–202; and Jack 'Scoup' Anderson (with James Boyd), *Oil*, Sidgwick and Jackson, London 1984.
58 Bina, *The Economics of the Oil Crisis*.
59 This was the basis for the famous editorial by *The Economist*, 7 July 1973:

> No one outside the US administration can prove whether this suspicion (that agreements with OPEC countries are the product of American corporate and government policy) is correct. It may be wholly incorrect. But even if it is, the fact that the suspicion exists inside the industry is significant, as is the guess, widespread in some places, as to at least one of the reasons for the administration's attitude. According to it, the Americans gave in to OPEC readily because they saw increased prices as a quick and easy way of slowing down the Japanese economy.

With the benefit of hindsight these suspicions were amply borne out.
60 See R. Ballance and S. Sinclair, *Collapse and Survival*, George Allen and Unwin, London 1983, pp. 160–77.
61 See Nore, 'Oil and the state', p. 87, n. 27.
62 Ballance and Sinclair, *Collapse and Survival*.
63 For details, see G. Luciani, *The Oil Companies and the Arab World*, Croom Helm, London 1984.
64 B. Levy, 'World oil marketing in transition', in *International Organization*, 36, 1982, pp. 113–33. The figures in the paragraph above come from this article.
65 Summarized in Roncaglia, *The International Oil Market*, pp. 66–7; see also Ballance and Sinclair, *Collapse and Survival*, pp. 166–77; and E. Shaffer, *The United States and the Control of World Oil*, Croom Helm, London 1983, pp. 192–209.
66 P. Armstrong, A. Glyn and J. Harrison, *Capitalism since World War II*, Fontana, London 1984, p. 310.
67 Here I am only concerned with the effects of the 1973–4 price increases; see chapter 6 for the increases of 1979. For two other analyses which

consider the general economic difficulties to be relatively independent of the oil crisis, see Lipietz, *The Enchanted World*; and Ernest Mandel, *The Second Slump*, Verso, London 1978. And for useful theoretical considerations, see Jose Luis Nicolini, 'The degree of monopoly, the macroeconomic balance and the international current account: the adjustment to the oil shocks', and Frank Vandenbroucke, 'Conflict in international economic policy and the world recession: a theoretical analysis', both in *Cambridge Journal of Economics*, 9, 1985, pp. 127–40 and 15–42.

68   Michael Bruno, 'World shocks, macroeconomic response, and the productivity puzzle', in R. C. O. Matthews (ed.), *Slower Growth in the Western World*, Heinemann, London 1982.

69   Richard Jackman, Comment in ibid.

70   William D. Nordhaus, 'Oil and Economic Performance in Industrial Countries', *Brookings Papers on Economic Activity*, 1980, no. 2.

71   See Bob Rowthorn, '"Late Capitalism"', in idem, *Capitalism, Conflict and Inflation*, pp. 113–15.

72   See Derek H. Aldcroft, *Full Employment*, Wheatsheaf, Brighton 1984, p. 67; and Chris Edwards, *The Fragmented World*, Methuen, London 1985, pp. 60–2 and 246–54.

73   Parboni, *The Dollar and Its Rivals*, p. 101.

74   Brian Tew, *The Evolution of the International Monetary System, 1945–1981*, Hutchinson, London 1982, p. 191.

75   On this, see Michel Aglietta, 'World capitalism in the eighties', in *New Left Review*, no. 136, 1982.

76   Tew, *The Evolution of the International Monetary System, 1945–1981*, p. 177.

77   See in particular the analysis of monetarism in Lipietz, *The Enchanted World*.

78   Fenn, *Oil and Diplomacy in the Twentieth Century*, p. 199, n. 41.

79   Krasner, *Structural Conflict*, p. 71.

80   Parboni, *The Dollar and Its Rivals*, pp. 49–56.

81   On these episodes, see S. Karlsson, *Oil and the World Order*, Berg, Leamington Spa 1986.

82   See IEA/OECD, *Energy Policies and Programmes of IEA Countries*, Paris 1984.

83   Armstrong et al., *Capitalism since World War II*, p. 312.

84   H. Magdoff and P. Sweezy, 'The U.S. Dollar, petrodollars, and U.S. imperialism', in idem, *The Deepening Crisis of U.S. Capitalism*, Monthly Review Press, New York, 1981, pp. 94–106.

85   See Terzian, *OPEC: The Inside Story*, p. 239.

86   See the summary in Roncaglia, *The International Oil Market*, pp. 86–96.

87   Charles A. Kupchan, *The Persian Gulf and the West*, Allen and Unwin, London 1987, p. 56.

88   Ibid., pp. 57–8.

89   The information on OPEC for the following account comes primarily from

Ian Skeet, *Opec: Twenty-five Years of Prices and Politics*, Cambridge University Press, Cambridge 1988, and Terzian, *OPEC: The Inside Story*.

90 Terzian, *OPEC: The Inside Story*, p. 110.

91 Ibid., p. 118. Skeet, *OPEC*, maintains that the smashing of the Saudi Tapline was 'directed against Saudi Arabia . . . and was not directly related to Libyan price aspirations nor support for OPEC, but the supplemental effect upon the oil market was dramatically helpful to Libya' (p. 60). It is hard to believe that Syria was unaware of this effect.

92 Ibid., p. 140.

93 Ibid., p. 212.

94 Ibid., pp. 216–17.

95 It has been suggested, particularly in the United States, that Saudi Arabian production decisions in late 1978 and 1979 were aimed at increasing instability in the market because of disapproval of US Middle East policy (see, for example, Kupchan, *The Persian Gulf and the West*). Skeet, *OPEC*, argues against this and concludes that 'an interpretation based on *ad hoc* perceptions and responses seems more persuasive and consistent with reality than one based on a rather elaborate structure of logic and contrived circumstantial evidence'(pp. 161–4). I tend to agree.

96 See Fred Halliday, *The Making of the Second Cold War*, Verso, London 1983, pp. 81–104 and *Cold War, Third World*, Hutchinson, London 1989, pp. 24–32 and 97–112. Chapter 5 will consider this issue in more detail.

97 The United States adopted two further strategies which I do not consider here: first, under the Reagan administration pressure was placed on international lending institutions not to finance oil developments which excluded participation by the oil companies; and second, the United States has built up a limited Strategic Petroleum Reserve.

98 On which, see P. Odell, *Oil and World Power*, Penguin, Harmondsworth 1983, pp. 231–8.

99 Venn, *Oil and Diplomacy in the Twentieth Century*, p. 148. See chapter 5 for a discussion of Japanese policy.

100 For further details, see ibid., pp. 148–51; and also M. Massarrat, 'The 'Energy Crisis': The struggle to redistribute surplus profit', in Nore and Turner (eds), *Oil and Class Struggle*, and Louis Turner, *The Oil Companies and the International System*, RIIA, George Allen and Unwin, London 1983, pp. 176–86.

101 Parboni, *The Dollar and Its Rivals*, p. 18.

102 See Keohane, *After Hegemony*, pp. 190–4 and 217–40.

103 Quoted in Shaffer, *The United States and the Control of World Oil*, p. 184 and more generally pp. 176–86.

104 Quoted in ibid., p. 180.

105 See Parboni, *The Dollar and Its Rivals*, pp. 52–6.

106 Roncaglia, *The International Oil Market*, pp. 121–8.

CHAPTER 5 ALLIES AND RIVALS: PETROPOLITICS IN EUROPE, JAPAN
AND THE SOVIET UNION

1   For further details on the Community's energy industries and policies, see Douglas Evans, *The Politics of Energy*, Macmillan, London 1975, and *Western Energy Policy*, Macmillan, London 1978; Stephen George, *Politics and Policy in the European Community*, Clarendon Press, Oxford 1985, pp. 107–14; and Dennis Swann, *The Economics of the Common Market*, Penguin, Harmondsworth 1984, pp. 244–62.

2   See Evans, *Western Energy Policy*.

3   Romano Prodi and Alberto Clo, 'Europe', in Raymond Vernon (ed.), *Daedalus: The Oil Crisis*, 104, 1975, p. 93.

4   Ibid., p. 98 and Evans, *The Politics of Energy*, p. 87.

5   Richard J. Samuels, *The Business of the Japanese State*, Cornell University Press, Ithaca 1987, p. 52.

6   See chapter 3 for a fuller discussion of this point.

7   Ibid., p. 61.

8   Lucas, *Western European Energy Policies*, p. 142.

9   Ibid., pp. 190–1.

10   Ibid., p. 254.

11   Prodi and Clo, 'Europe', p. 103.

12   This paragraph draws directly on P. Gibbon and S. Bromley, '"From an institution to a business"? Changes in the British coal industry, 1985–9', in *Economy and Society*, 19, 1990, pp. 56–94.

13   See Phil Wright, 'Energy Policy and Coal – The View From the European Community', mimeo. See also D. Feickert, *International Labour Reports* Sept/Oct, no. 29, 1988. World coal trade expanded from 304.7mt in 1984 to 340.9mt in 1987. Of this total, steam/coal trade amounted to 141.1mt in 1987, with a further 100mt expected to be added over the next decade.

14   The *Jahrhundertvertrag* obliges the electricity utilities to buy at least 45mt of German deep-mined coal a year until 1995, while the *Kohlepfennig* is a levy on electricity prices – currently 7.5 per cent – to cover the difference between the prices of domestic and internationally traded coal.

15   For much greater detail on British policy and full references, see S. Bromley, *The State, Capital and the Oil Industry*, Ph. D. diss., University of Cambridge, 1987; and from an extensive literature the extremely valuable studies of P. Cameron, *Property Rights and Sovereign Rights*, Academic Press, London 1983; William Keegan, *Britain without Oil*, Penguin, Harmondsworth 1984, and A. Porter, M. Spence and R. Thomspon, *The Energy Fix*, Pluto Press, London 1985.

16   Again, for a full account and references, see Bromley, *The State, Capital and the Oil Industry*. For recent details, see the annual *OECD Economic Surveys*, Paris.

17   David Calleo, *The German Problem Reconsidered*, Cambridge University Press, Cambridge 1978, pp. 186–187.

18 The following sentences draw directly on S. Bromley and J. Rosenberg, 'After Exterminism', in *New Left Review*, no. 168, pp. 84–5. And for important background discussion, see D. Johnstone, *The Politics of Euromissiles*, Verso, London 1984; and O. MacDonald, 'New directions in Soviety "Westpolitik"?', in *Labour Focus on Eastern Europe*, May 1986.

19 Stephen Woolcock, *Western Policies on East–West Trade*, RIIA/Routledge & Kegan Paul, London 1982. This paragraph draws directly on Woolcock's cogent analysis.

20 See R. Lieber, 'Energy policy and national security: Invisible hand or guiding hand?', in K. Oye, R. Lieber and D. Rothschild (eds), *Eagle Defiant*, Little, Brown, Boston 1983, pp. 176–9. US thinking was as follows: poor Soviet harvests implied greater grain imports which in turn required foreign exchange. But as oil exports were set to decline, hard currency would be scarce. Without offsetting increases in gas exports, therefore, the Soviets would have to reduce capital and consumer goods imports in order to pay for food. This would have two benefits: first, US exports to the region were predominantly grain whereas those of Western Europe were mainly industrial; and second, reduced imports of advanced goods would weaken the Soviet economy. See Woolcock, *Western Policies on East–West Trade*, p. 68.

21 According to Philip Hanson, *Western Economic Statecraft in East–West Relations*, RIIA/Routledge and Kegan Paul, London 1988, p. 24, 35 per cent was the upper limit set by 'an informal understanding within NATO'.

22 See Jonathan P. Stern, 'East–West energy trade', in R. Belgrave (ed.), *Energy – Two Decades of Crisis*, Heinemann, London 1983, p. 77.

23 Thus, Hanson *Western Economic Statecraft in East–West Relations*, notes that: 'In Washington the general perspective was one of economic "containment" or, to put it more bluntly, economic warfare' (p. 47).

24 *The Economist*, 22 April 1989.

25 Jeffrey Pryce, 'The Atlantic Alliance and the Yom Kippur War', in *The Cambridge Review of International Studies* Vol. 1 no. 1 1986, p. 28.

26 Elizabeth Heneghan, 'Acrimony and alliance: European political cooperation and the Middle East', in *The Cambridge Review of International Studies*, 1, 1986, p. 34.

27 See R. Lieber, 'Cohesion and disruption in the Western alliance', in D. Yergin and M. Hillenbrand (eds), *Global Insecurity*, Penguin, Harmondsworth 1982.

28 See the relevant tables in the World Bank's *World Development Report, 1989* and the IMF's *World Economic Outlook, 1989*. Also see Fred Lawson, 'The Reagan administration in the Middle East', in *Middle East Reports*, 14, 1984.

29 See chapter 6 for a further discussion of this point.

30 Charles A. Kupchan, *The Persian Gulf and the West*, Allen and Unwin, London 1987, p. 197.

31 For a discussion, see Jochen Hippler, 'NATO goes to the Persian Gulf',

in *Middle East Report*, no. 155, Nov./Dec. 1988. The shift in European opinion can be seen in the alteration of the editorial line in the *Financial Times* between October and November 1987. At first the FT argued that US attacks on Iranian oil platforms were potentially escalatory, that the Soviet proposal for joint action through the UN Security Council was 'worthy of serious consideration', and that the European allies should remind the United States 'of the dangers of going it alone' (20 October 1987). Later a more benign view was taken of US actions and while the Soviet proposal was still considered to be 'worth exploring, especially if it is the price of Soviet cooperation in increasing the pressure on Iran', attention was drawn to the possibility of 'an insidious attempt to establish a Soviet veto over Western naval activity' in the Gulf through the Security Council (30 November 1987).

32　Christopher Thorne, *The Far Eastern War*, Unwin, London 1986, p. 25.

33　See chapter 2. See also Paul Kennedy, 'Japanese strategic decisions, 1939–45', in idem, *Strategy and Diplomacy, 1870–1945*, George Allen and Unwin, London 1983.

34　Michael Schaller, The *American Occupation of Japan*, Oxford University Press, Oxford 1985, p. 77. It is in general fruitless to debate whether strategic or economic considerations dominated US policy-making in this regard since the two were utterly imbricated. The fundamental point was both simple and widely appreciated: rebuilding a capitalist world economy under US leadership and the 'containment' of the Soviet Union were indissolubly linked not because the Soviets were bent on expansion or posed a military threat but because 'Soviet support for revolutionary movements – whether rhetorical or material – could transform the political balance of Europe, the Middle East, and Asia' (ibid, p. 78). For other treatments which avoid the sterile and superficial duality of 'economic' versus 'strategic' accounts, see the pioneering essay by Jon Halliday, 'Captialism and socialism in East Asia', in *New Left Review*, no. 124, 1980. See also Jon Halliday and Gavan McCormack, *Japanese Imperialism Today*, Penguin, Harmondsworth 1973; and Howard Schonberger, 'U.S. policy in post-war Japan: The retreat from liberalism', in *Science and Society*, 46, 1982, pp. 39–59. And for a realist treatment which ignores Japan's role in the stabilization of East Asia, preferring to argue that Japan was merely a mercantilist free-rider on the open international economy provided by the United States, see Michael Mandelbaum, 'The international economic order and Japan, 1945–1985', in idem, *The Fate of Nations*, Cambridge University Press, Cambridge 1988.

35　Schaller, *The American Occupation of Japan*, pp. 279 and 288.

36　For an important discussion of this vexed question, see Andrea Boltho, 'Was Japan's industrial policy successful?', in *Cambridge Journal of Economics* 9, 1985, pp. 187–201. As Boltho explains, Italy, for example, faced a technological gap comparable to Japan, also existed in a world market characterized by buoyant demand and was specialized much more than

Japan in branches of manufacturing where demand grew rapidly. Comparing Japan's postwar growth either with its pre-war performance or with other countries, therefore, demonstrates an 'unexplained residual' in Japanese growth not accounted for by export demand.

37 See the original essay by Bruce Cumings, 'The origins and development of the Northeast Asian political economy: industrial sectors, product cycles and political consequences', in *International Organization* 38, 1984, pp. 2–3.

38 Michio Morishima, *Why Has Japan 'Succeeded'?*, Cambridge University Press, Cambridge 1982, p. 190.

39 On Japanese oil, see Samuels, *The Business of the Japanese State*; T. Murakami, 'The remarkable adaptation of Japan's economy', and J. Watanuki, 'Japanese society and the limits to growth', in both in Yergin and Hillenbrand (eds), *Global Insecurity*; P. Odell, *Oil and World Power*, Penguin, Harmondsworth 1983; and Yoshi Tsurumi, 'Japan', in Vernon (ed.), *The Oil Crisis*.

40 Samuels, *The Business of the Japanese State*, p. 184.

41 Quoted in Schaller, *The American Occupation of Japan*, p. 179.

42 Ibid., p. 191.

43 Ibid., p. 205.

44 Raymond Vernon, *Two Hungry Giants*, Harvard University Press, Cambridge, Mass. 1983, p. 2. The exact figures are as follows: in 1979 the United States and Japan accounted for 46 per cent of the non-communist world's oil and 39 per cent of its imports.

45 Tsurumi, 'Japan', p. 124.

46 Ibid. Note also Cumings's comment that 'in the postwar world economy Japan resembles a sector as much as a nation-state', in 'The origins and development of the Northeast Asian political economy', p. 17.

47 Vernon, *Two Hungry Giants*, p. 97.

48 Samuels, *The Business of the Japanese State*, p. 215.

49 See R. Dore, 'Energy conservation in Japanese industry', in Belgrave (ed.), *Energy - Two Decades of Crisis*.

50 William Feeney, 'Chinese policy in multilateral financial institutions', in Samuel S. Kim (ed.), *China and the World*, Westview Press, Boulder 1984, p. 276.

51 B. Garrett, 'China policy and the constraints of triangular logic', in K. Oye et al. (eds.), *Eagle Defiant*, p. 238. See also Bruce Cumings, 'The political economy of China's turn outward', in Kim (ed.), *China and the World*.

52 Laura Newby, *Sino-Japanese Relations*, RIIA/Routledge and Kegan Paul, London 1988, p. 8. For further analysis of the Siberian connection, see the discussion below of the Soviet Union.

53 Ibid., p. 26. This refers to the position in the late 1980s.

54 See Michel Chossudovsky, *Towards Capitalist Restoration? Chinese Socialism after Mao*, Macmillan, London 1986, pp. 165–71.

55 Ryutaro Komiya and Motoshige Itoch, 'Japan's international trade and

trade policy, 1955–1984', in Takashi Inoguchi and Daniel I. Okimoto (eds), *The Political Economy of Japan, Vol. 2 The Changing International Context*, Standford University Press, Standford 1988, p. 199.

56 Rod Steven, 'The high yen crisis in Japan', in *Capital and Class*, Spring 1988, pp. 84 and 87. An excellent survey.

57 Ibid., p. 102.

58 Paul Maidment, 'The Yen Block', a survey in *The Economist*, 15 July 1989, p. 15. Note also that US pressure on trade began to pay off by the late 1980s: between 1985 and 1988 the volume of Japanese imports of manufactures rose by 80 per cent, US manufactured goods exports were up 79 per cent and total US exports were up 67 per cent.

59 See Jerry F. Hough, *Opening up the Soviet Economy*, Brookings Institution, Washington, D.C. 1988, p. 10, n. 4, citing, K. Ohmae, 'Japan's trade failure', in the *Wall Street Journal*, 1 April 1987. For a useful general discussion, see Robert Gilpin (with the assistance of Jean M. Gilpin), *The Political Economy of International Relations*, Princeton University Press, Princeton 1987, pp. 328–40. Note also that while Japan's financial and monetary position increased rapidly in the 1980s, taking its banks and financial institutions to the position of world leaders, the yen's share of international bond issues rose from 1.6 per cent in 1980 to 8.9 per cent in 1988, its share of official currency reserves increased from 4.3 to 8.0 per cent and the proportion of Japan's trade denominated in yen expanded from 2.4 to 10.6 per cent, and yet the dollar remained predominant, with 60 per cent of bond issues and 64 per cent of reserves. The yen's position can also be compared to that of the West German D-mark which accounted for 17 per cent of currency reserves in 1988.

60 See the valuable discussion in Reinhard Drifte, *Japan's Foreign Policy*, RIIA/Routledge, London 1990. Drifte argues that:

> Although it is still firmly integrated into the alliance with the United States, Japan is increasingly supplementing US military power in Asia, and the United States is encouraging it to play a more active political role in promoting the process of Asia's economic and political development. . . . Japan has become indispensable for the management of the global economic system and the maintenance of international security. For the United States, in particular, Japan has become the most important single country in supporting its super-power status. (pp. 62 and 103).

61 See M. Tanzer and S. Zorn, *Energy Update*, Monthly Review Press, New York 1985, pp. 106–11; and Marshall I. Goldman, *The Enigma of Soviet Petroleum*, George Allen and Unwin, London 1980, p. 13–31. See also Alec Nove, *An Economic History of the U.S.S.R.*, Penguin, Harmondsworth 1982, pp. 14, 101, 226, 230–1, 273, 293, 342, 354–5 and 376–7.

62 Goldman, *The Enigma of Soviet Petroleum*, p. 35.·

63 Ibid., p. 60.

64 Quoted in ibid., p. 70

65 Ibid., p. 70.

66 Tanzer and Zorn, *Energy Update*, p. 111.
67 See P. James, *The Future of Coal*, Macmillan, London 1984, p. 126.
68 See R. Pipes, *Survival is not Enough*, Simon and Schuster, New York 1984, p. 184, citing Richard Portes, *Deficits and Detente*, New York 1983.
69 Cited in David R. Marples, *Chernobyl & Nuclear Power in the USSR*, Macmillan, London 1986, p. 47.
70 See ibid., pp. 37–93 for a thorough discussion.
71 Ibid., p. 51.
72 Quoted in ibid., p. 177.
73 See Paul Dibb, *The Soviet Union: The Incomplete Superpower*, Macmillan/ IISS, London 1988, for a compelling discussion of the Siberian problem. The Soviet strategic worries have been noted by Jonathan Steele, *The Limits of Soviet Power*, Penguin, Harmondsworth 1984, p. 139, in his discussion of the Sino-Soviet border dispute: 'Two of Siberia's major cities, Vladivostok and Khabarovsk, lay in the region under dispute. In addition, the Trans-Siberian Railroad, which is the only lateral land route from central Russia to the Pacific, runs within twenty miles of the Chinese border for most of the last eight hundred miles of its length. Any cut in the line would block overland supplies to eastern Siberia and interrupt the main source of oil to the Pacific Fleet'.
74 Ibid., p. 57.
75 See Nove, *An Economic History of the U.S.S.R.*, p. 379.
76 See Abel Aganbegyan, *The Challenge: Economics of Perestroika*, Hutchinson, London 1988, p. 103. Aganbegyan's figures suggest that the potential for savings are indeed immense: it takes 'two to two and a half times less resources to save one tonne of fuel in the USSR by economising than it does to mine one extra tonne. . . . [And] prices for fuel and raw materials . . . are usually two to three times lower than the world price' (p. 72).
77 Anders Asland, *Gorbachev's Struggle for Economic Reform*, Pinter, London 1989, p. 72.
78 James Sherr, *Soviet Power: The Continuing Challenge*, Macmillan/RUSI, London 1987, p. 25.
79 *BP Statistical Review of World Energy*, pp. 2 and 20.
80 For a general discussion of considerable insight, see Hough, *Opening up the Soviet Economy*, especially ch. 4.
81 Quoted as an epigraph to Steele's chapter on Soviet policy in the Middle East, *The Limits of Soviet Power*, p. 179.
82 Quoted in ibid., p. 191.
83 Shoshana Klebanoff, *Middle East Oil and U.S. Foreign Policy*, Praeger, New York 1974, p. 167.
84 Fortunately there is not the space to go into this here in any detail, but see Fred Halliday, 'The Middle East, Afghanistan and the Gulf in the Soviet perception', in Sherr, *Soviet Power: The Continuing Challenge*; in Michael MccGwire, *Military Objectives in Soviet Foreign Policy*, Brookings, Washington, D.C. 1987, pp. 183–210; and Thomas L. McNaugher, *Arms*

and Oil: U.S. Military Strategy in the Persian Gulf, Brookings, Washington, D.C. 1985, ch. 2. Briefly, the drive for warm water ports occurred from the ninth to the eighteenth centuries and concerned the northern shores of the Black Sea; domestic problems in Soviet Central Asia are real but have not involved (as yet) the spread of radical Islam from Iran; and the supposed need for oil is simply absurd: not only will the Soviet Union be a net exporter of petroleum for the foreseeable future but arguments that it would invade the Middle East ignore two huge costs: first, there are immense political and military problems of controlling the region; and second, it did not require the Carter Doctrine to alert the Soviet Union to the fact that the West has a vital interest in the region. At worst, given serious instability in Iran and with a US intervention in the south, the Soviet Union might in some circumstances invade northern Iran. On the indigenous character of the region's revolutions and the Soviet role, see Fred Halliday, *The Threat from the East?*, Penguin, Harmondsworth 1982.

85 McNaugher, *Arms and Oil*, p. 23.
86 Steele, *The Limits of Soviet Power*, p. 201; and see especially MccGwire, *Military Objectives in Soviet Foreign Policy*. McNaugher describes the Soviet naval buildup in the Indian Ocean as 'reactive', *Arms and Oil*, p. 45.
87 Efraim Karsh, *The Soviet Union and Syria*, RIIA/Routledge and Kegan Paul, London 1988, p. 3.
88 See Robin Edmonds, *Soviet Foreign Policy, 1962–1973*, Oxford University Press, Oxford 1975; and Fred Halliday, *Cold War, Third World*, Hutchinson, London 1989, ch. 4. Some commentators also detect a dynamic to Soviet policy in the Third World during this period deriving from the Soviet Communist Party's rivalry with the Chinese Communist party for influence among revolutionary movements.
89 See Dina Rome Spechler, 'The politics of intervention: The Soviet union and the crisis in Lebanon', in *Studies in Comparative Communism*, 20, Summer 1987. Spechler claims that prior to 1973 the politburo oscillated between a strong pro-Arab, anti-US stance in the region and subordinating regional concerns to wider superpower, East–West relations.
90 Karsh, *The Soviet Union and Syria*, p. 99; see also Klebanoff, *Middle East Oil and U.S. Foreign Policy*, chs 8 and 9.
91 See Samir Amin, *The Arab Nation*, Zed Press, London 1978. See also Ahmad N. Azim, 'Egypt: The origins and development of a neo-colonial state'; Fred H. Lawson, 'Class politics and state power in Ba'thi Syria'; and Joe Stork, 'Class, state and politics in Iraq' – all in Berch Berberoglu (ed.), *Power and Stability in the Middle East*, Zed Press, London 1989.
92 Spechler, 'The politics of intervention', pp. 137–8. See further the analyses in Halliday, *Cold War, Third World*. Jerry F. Hough, *The Struggle for the Third World*, Brookings, Washington, D.C. 1986; and Stephen Shenfied, *The Nuclear Predicament*, RIIA/Routledge and Kegan Paul, London 1987.

93   See the conclusion to chapter 6 for further discussion of this.
94   The former Soviet doctrine of the 'correlation of forces', however crass and ritualistic, at least had the merit of attempting to grasp the structural dimensions of the balance of social forces both between and within states which laid down many of the parameters of international politics. The flatulence of much contemporary Soviet rhetoric comes from its refusal to recognize that the Soviet inability or unwillingness to challenge the West economically and politically will not be reciprocated by a comparable Western accommodation to communism. As Fred Halliday has recently noted, Soviet 'new thinking' meant that for the West 'the 'reversibility' of communism was now on the agenda' (*Cold War, Third World*, p. 134). Or, to recycle a phrase from 'the great communicator', Ronald Reagan, you can run, but you can't hide.
95   See Dibb, *The Soviet Union: The Incomplete Superpower* for a good treatment of this point.
96   Newby, *Sino-Japanese Relations*, p. 69.
97   Dibb, *The Soviet Union: The Incomplete Superpower*, p. 20.
98   *Newby, Sino-Japanese Relations*, p. 81.
99   Zhao Ziyang's phrase quoted in ibid., p. 76.
100 See C. Coker, 'Eastern Europe and the Middle East: The forgotten dimension of Soviet policy', in Robert Cassen (ed.), *Soviet Interests in the Third World*, RIIA/Sage, London 1985.

CHAPTER 6 US STRATEGY IN THE 1980s

1   Robert Gilpin, *War and Change in World Politics*, Cambridge University Press, Cambridge 1981.
2   See I. Wallerstein, *The Politics of the World-Economy*, Cambridge University Press, Cambridge 1984; Paul Kennedy, *The Rise and Fall of Great Powers*, Unwin Hyman, London 1988; and also J. O'Loughlin, 'World-power competition and local conflicts in the Third World', in R. Johnston and P. Taylor (eds), *A World Crisis?*, Basil Blackwell, Oxford 1986.
3   See chapter 2.
4   For further discussion, see K. Oye, 'International systems structure and American foreign policy', in K. Oye, R. Lieber and D. Rothchild (eds), *Eagle Defiant*, Little, Brown, Boston 1983 and Riccardo Parboni, 'The dollar weapon: From Nixon to Reagan', in *New Left Review*, no. 158, 1986.
5   See the discussion in Fred Halliday, *The Threat from the East?*, Penguin, Harmondsworth 1982.
6   See Isaac Deutscher, *The Great Contest*, Oxford University Press, London 1960.
7   The following remarks have been much influenced by Michael Cox, 'The Cold War as a system' in *Critique*, no. 17, 1986; Fred Halliday, *The Making of the Second Cold War*, Verso, London, 1983; Riccardo Parboni,

The Dollar and Its Rivals, Verso, London 1981; and Kees van der Pijl, *The Making of an Atlantic Ruling Class*, Verso, London 1984.

8   Halliday, *The Making of the Second Cold War*, p. 224.

9   Mike Davis, 'The political economy of late imperial America' and 'Reaganomics' Magical Mystery Tour' in *New Left Review*, nos 143 and 149, 1984 and 1985, respectively.

10 Oye, 'International systems structure and American foreign policy', p. 13. It is perhaps no accident that 'the voluble Laffer is so persuasive that he managed to lure the faculty of the University of Chicago into making him a tenured professor under the impression that he had a Ph.D (though, at the time, he did not)'; see Garry Wills, *Reagan's America*, Doubleday, New York 1987, p. 448, n. 13.

11 Alain Lipietz, 'How monetarism choked Third World industrialization', in *New Left Review*, no. 145, 1984.

12 See the analysis by T. Amott, 'The politics of Reaganomics', in E. Nell (ed.), *Free Market Conservatism*, Macmillan, London 1984.

13 Robert Heilbroner and Peter Bernstein, *The Debt and the Deficit*, W. W. Norton, New York 1989, p. 27. It is perhaps worth noting that the origins of and connections between the twin deficits are the subject of some debate. First, the origins of the trade deficit. Some argue that approximately two-thirds of the trade deficit is due to the dollar appreciation in the early 1980s, and the rest is due to buoyant growth in the United States as compared with elsewhere. Superimposed on this, and offsetting favourable movements in the real price of oil, are such factors as the debt crisis (reducing Third World imports) and the continuing rapid growth of the NICs (competitors in home and overseas markets). Others argue that up to two-thirds of the present trade gap is *unrelated* to the dollar's rise, it rather reflects the cumulative difference between the growth of demand in the United States and other industrialized countries, a $60 billion trade deterioration with the Asian NICs and the debt crisis. It is generally agreed that the proximate cause of the overvalued dollar was the high interest rate policy of the dederal Reserve. The first account, therefore, sees only a weak connection between the deficits (the widening fiscal deficit plays a small role in generating the trade deficit through increasing growth and maintaining the pressure on interest rates), while the second sees a stronger connection as the fiscal deficit generated the growth which largely accounts for the trade deficit. Second, the fiscal deficit. Those who argue that the fiscal deficit is a structural (full-employment) one tend to argue a strong connection between the deficits. Ronald McKinnon thus argued that the structural deficit by channelling savings away from investment would 'continue to show up as a large trade deficit of the same order of magnitude – whether the dollar is devalued or not' (*Financial Times*, 9 December 1987). On the other hand, others argue – in my view in line with the run of evidence – that the structural deficit is relatively unimportant and that the deficits are both a cyclical (recession) phenomenon and a result rather than

a cause of high interest rates (as high interest rates inflate net interest repayments). As Cornwall argued: 'The deficits recently being experienced throughout the OECD were reflecting the poor performance of the private sector, partly induced by tight monetary policies . . . Related, the high interest rates should be seen as a cause of the large deficits and not vice versa'; see ibid., *The Conditions for Economic Recovery*, Martin Robertson, Oxford 1983, p. 211. The trade deficit is thus not umbilically linked to the fiscal deficit.

14 Heilbroner and Bernstein, *The Debt and the Deficit*, p. 26.
15 Davis, 'The political economy of late imperial America', p. 36.
16 Parboni, 'The dollar weapon: From Nixon to Reagan', p. 15.
17 T. Edsall, *The New Politics of Inequality*, Pantheon New York 1984; and on the international rise of rentier interests, see A. Bhaduri and J. Steindl, 'The rise of monetarism as a social doctrine', *Thames Papers in Political Economy*, 1983.
18 Davis, 'Reaganomics' Magical Mystery Tour', pp. 57–8.
19 See Andre Gunder Frank, *The European Challenge*, Spokesman, Nottingham 1983, pp. 19–54.
20 Davis, 'Reagonomics' Magical Mystery Tour'.
21 Parboni, 'The dollar weapon: From Nixon to Reagan', pp. 13–14.
22 For an analysis along these lines, see John Palmer, *Europe without America?*, Oxford University Press, Oxford 1987.
23 IEA, *Energy Policies and Programmes of IEA Countries: 1984 Review*, OECD, Paris 1985.
24 Calculated from ibid., pp. 55 and 61.
25 See P. Odell, 'Draining the World of Energy', in Johnston and Taylor (eds), *A World Crisis?*.
26 Reported in R. Bending and R. Eden, *UK Energy*, Cambridge University Press, Cambridge 1984, pp. 24–7.
27 See P. Odell and K. Rosing, *The Future of Oil*, Kogan Page, London 1983; and *World Development Report 1983*, World Bank.
28 P. James, *The Future of Coal*, Macmillan, London 1984; and for further discussion of world coal, see Richard L. Gordon, *World Coal*, Cambridge University Press, Cambridge 1987.
29 Odell, 'Draining the World of Energy', p. 76.
30 P. Odell, *Oil and World Power*, Penguin, Harmondsworth 1983, pp. 264 and 231–70.
31 See the work of M. Tanzer and S. Zorn, 'OPEC's decade: Has it made a difference', in *Middle East Report*, 14, 1984 and in more detail, *Energy Update*, Monthly Review Press, New York 1985. Much additional material in the following sections comes from the economic press, including *Business Week, The Economist* and the *Financial Times*.
32 See R. Ballance and S. Sinclair, *Collapse and Survival*, George Allen and Unwin, London 1983, pp. 160–77; and the economic press.
33 See Charles J. DiBona, 'Long-term trends in oil markets', in Edward R.

Fried and Nanette M. Blandin (eds), *Oil and America's Security*, Brookings, Washington, D.C. 1988.

34 Quoted in Noam Chomsky, *The Fateful Triangle*, Pluto Press, London 1983, p. 17.

35 Rosemarie Said Zahlan, *The Making of the Modern Gulf States*, Unwin Hyman, London 1989, p. 124.

36 See Jonathan Marshall, 'Saudi Arabia and the Reagan Doctrine', in *Middle East Report*, no. 155, 1988.

37 J. Stork, 'Ten years after', in *Middle East Report*, 14, 1984, p. 4.

38 Charles A. Kupchan, *The Persian Gulf and the West*, Allen and Unwin, London 1987, p. 67.

39 See Halliday, *The Threat from the East?*.

40 Kupchan, *The Persian Gulf and the West*, p. 213.

41 For some of the background, see Michael T. Klare, *Beyond the 'Vietnam Syndrome'*, Institute for Policy Studies, Washington 1981.

42 For useful accounts, see M. Klare, 'Intervention and the nuclear firebreak in the Middle East'; F. Lawson, 'The Reagan administration in the Middle East'; and M. Wenger, 'The central command: Getting to the war on time'; all in *Middle East Report*, 14, 1984; as well as B. Rubin, 'The Reagan administration and the Middle East', in Oye, Lieber and Rothchild (eds), *Eagle Defiant*. The irrelevance of Soviet 'subversion' in the Gulf Peninsula has been underscored by Thomas L. McNaugher, *Arms and Oil: U.S. Military Strategy and the Persian Gulf*, Brookings Institution, Washington, D.C. 1985, p. 106: 'The prospects for communist-inspired subversion are fairly low. More likely would be shifts to the right (toward more conservative Islamic institutions), toward the center (a coup based on middle-class technocrats), or toward a military regime likely to have military ties to the West'.

43 Kupchan, *The Persian Gulf and the West*, pp. 146–8.

44 For discussion of the Gulf Co-operation Council, see Nancy C. Troxler, 'The Gulf Co-operation Council: The emergence of an institution', in *Millennium*, 16, and more generally Muhammad Rumaihi, *Beyond Oil*, Al Saqi Books, London 1986.

45 See J. Cypher, 'Monetarism, militarism and markets: Reagan's response to the structural crisis' in *Middle East Report*, 14, 1984.

46 See Lawrence Lifschultz, 'From the U-2 to the P-3: The US–Pakistan relationship', in *New Left Review*, no. 159, 1986 pp. 71–8; and Jamal Rashid, 'Pakistan and the Central Command', in *Middle East Report*, 16, 1986.

47 For a careful discussion of this point, see Mohammed Ayoob, 'The Iran–Iraq War and regional security in the Persian Gulf', in *Alternatives* 1985.

48 See Fred Halliday, 'A good revolutionary pause', in *New Statesman and Society*, 29 July 1988; and Joe Stork, 'Class, state and politics in Iraq', in

Berch Berberoglu (ed.), *Power and Stability in the Middle East*, Zed Press, London 1989, pp. 49–52.

49   A point stressed in the important article by George Joffe, 'Stars and Stripes over the Gulf', in *Journal of European Nuclear Disarmament*, no. 30 1987.

50   As is alleged by Dilip Hiro; see Malcolm Yapp, 'Slaughter in the Gulf' (reviewing Dilip Hiro, *The Longest War* and Efraim Karsh ed., *The Iran–Iraq War*), in *Times Literary Supplement* 4 and 10 August 1989.

51   Said Zahlan, *The Making of the Modern Gulf States*, p. 158.

52   Yapp, 'Slaughter in the Gulf'.

53   Ian Skeet, *OPEC: Twenty-five years of prices and politics*, Cambridge University Press, Cambridge 1988, p. 184.

54   Ibid., pp. 185–6.

55   See ibid., pp. 186–93.

56   For a detailed discussion of the financial policies of OPEC's members, see Richard Mattione, *OPEC's Investments and the International Financial System*, Brookings, Washington, D.C. 1985.

57   Though it was superficially similar to the policy adopted in 1986, by 1988 Saudi Arabia had adopted a new strategy. This time the Kingdom was forcefully demonstrating that it would no longer play the role of a swing producer.

58   For a detailed analytical treatment , see R. Mabro, R. Bacon, M. Chadwick, M. Halliwell and D. Long, *The Market for North Sea Crude*, Oxford University Press, Oxford 1986.

59   Ibid., p. 241.

60   Samir Amin, *The Arab Economy Today*, Zed Press, London 1982, pp. 78–80.

61   Fred Halliday, '1967 and the consequences of catastrophe', in *Middle East Report*, no. 146, 1987, p. 4.

62   See also Stork, 'Ten years after' and Y. A. Sayigh, 'The Arab oil economy', in Talal Asad and Roger Owen (eds), *The Middle East*, Macmillan, London 1983.

63   Amin, *The Arab Economy Today*, p. 41.

64   See Mattione, *OPEC's Investments in the International Financial System*, p. 83.

65   See Robert H. Ballance, *International Industry and Business*, Allen and Unwin, London 1987, pp. 172–3. Sabic is currently 70 per cent state-owned, though there are increasing calls for some privatization as a Saudi capitalist class begins to consolidate itself. See generally Ghassan Salame, 'Political power and the Saudi state', in Berberoglu (ed.), *Power and Stability in the Middle East*.

66   Jamie Simpson, 'Compensatory trade, defence procurement and industrial policy' (London School of Economics, mimeo.). I am very grateful to the author for allowing me to see this as yet unpublished work, and my comments in this paragraph draw directly on his arguments.

67   Of course, Marxists too are capable of similar claims about the overall loss

of US power in the world oil industry and relatedly in hegemonic leadership, see for example Michael Tanzer, 'Growing instability in the international oil industry', in Arthur MacEwan and William K. Tabb (eds), *Instability and Change in the World Economy*, Monthly Review Press, New York 1989.

68 Of course, congressional controls complicate the ability of the executive to exploit this advantage.

69 Thus, Karen Dawisha concludes her survey 'The correlation of forces and Soviet policy in the Middle East' as follows: 'Military aims and instruments have . . . come to the forefront of Soviet policy in the Middle East. While undeniably establishing the USSR as a superpower in the area, the promotion of Soviet military power has emphasized the failings of Moscow's political and ideological offensive, and itself has brought with it problems which the Soviet leaders have yet to resolve'; see Adeed and Karen Dawisha (eds), *The Soviet Union in the Middle East*, Heinemann/RIIA, London 1982, p. 164.

70 Susan Strange, *States and Markets*, Pinter, London 1988, pp. 235–9. Given the institutional dominance of North American emphases within the international relations literature and academic community, I see little evidence for Strange's judgement that the view here propounded 'is held by a too-silent majority of non-American scholars'. If true, then it's a veritable conspiracy of silence. Unfortunately, Susan Strange's bold attempt to raise questions of structural power has been met by deafening silence on both sides of the Atlantic. A significant exception to this is the article by Stephen Gill, 'American hegemony: Its limits and prospects in the Reagan era', in *Millennium: Journal of International Studies*, 15, 1986, pp. 311–36; see also Stephen Gill and David Law, *The Global Political-Economy*, Harvester, London 1988; and for a brief if untheorized recognition of this point, see John A. Hall and G. John Ikenberry, *The State*, Open University Press, Milton Keynes 1989, pp. 83–92.

Index